REAL FUNCTION
ALGEBRAS

MONOGRAPHS AND TEXTBOOKS IN
PURE AND APPLIED MATHEMATICS

51. *A. C. M. van Rooij*, Non-Archimedean Functional Analysis (1978)
52. *L. Corwin and R. Szczarba*, Calculus in Vector Spaces (1979)
53. *C. Sadosky*, Interpolation of Operators and Singular Integrals (1979)
54. *J. Cronin*, Differential Equations (1980)
55. *C. W. Groetsch*, Elements of Applicable Functional Analysis (1980)
56. *I. Vaisman*, Foundations of Three-Dimensional Euclidean Geometry (1980)
57. *H. I. Freedan*, Deterministic Mathematical Models in Population Ecology (1980)
58. *S. B. Chae*, Lebesgue Integration (1980)
59. *C. S. Rees et al.*, Theory and Applications of Fourier Analysis (1981)
60. *L. Nachbin*, Introduction to Functional Analysis (R. M. Aron, translator) (1981)
61. *G. Orzech and M. Orzech*, Plane Algebraic Curves (1981)
62. *R. Johnsonbaugh and W. E. Pfaffenberger*, Foundations of Mathematical Analysis (1981)
63. *W. L. Voxman and R. H. Goetschel*, Advanced Calculus (1981)
64. *L. J. Corwin and R. H. Szczarba*, Multivariable Calculus (1982)
65. *V. I. Istrățescu*, Introduction to Linear Operator Theory (1981)
66. *R. D. Järvinen*, Finite and Infinite Dimensional Linear Spaces (1981)
67. *J. K. Beem and P. E. Ehrlich*, Global Lorentzian Geometry (1981)
68. *D. L. Armacost*, The Structure of Locally Compact Abelian Groups (1981)
69. *J. W. Brewer and M. K. Smith, eds.*, Emily Noether: A Tribute (1981)
70. *K. H. Kim*, Boolean Matrix Theory and Applications (1982)
71. *T. W. Wieting*, The Mathematical Theory of Chromatic Plane Ornaments (1982)
72. *D. B. Gauld*, Differential Topology (1982)
73. *R. L. Faber*, Foundations of Euclidean and Non-Euclidean Geometry (1983)
74. *M. Carmeli*, Statistical Theory and Random Matrices (1983)
75. *J. H. Carruth et al.*, The Theory of Topological Semigroups (1983)
76. *R. L. Faber*, Differential Geometry and Relativity Theory (1983)
77. *S. Barnett*, Polynomials and Linear Control Systems (1983)
78. *G. Karpilovsky*, Commutative Group Algebras (1983)
79. *F. Van Oystaeyen and A. Verschoren*, Relative Invariants of Rings (1983)
80. *I. Vaisman*, A First Course in Differential Geometry (1984)
81. *G. W. Swan*, Applications of Optimal Control Theory in Biomedicine (1984)
82. *T. Petrie and J. D. Randall*, Transformation Groups on Manifolds (1984)
83. *K. Goebel and S. Reich*, Uniform Convexity, Hyperbolic Geometry, and Nonexpansive Mappings (1984)
84. *T. Albu and C. Năstăsescu*, Relative Finiteness in Module Theory (1984)
85. *K. Hrbacek and T. Jech*, Introduction to Set Theory: Second Edition (1984)
86. *F. Van Oystaeyen and A. Verschoren*, Relative Invariants of Rings (1984)
87. *B. R. McDonald*, Linear Algebra Over Commutative Rings (1984)
88. *M. Namba*, Geometry of Projective Algebraic Curves (1984)
89. *G. F. Webb*, Theory of Nonlinear Age-Dependent Population Dynamics (1985)
90. *M. R. Bremner et al.*, Tables of Dominant Weight Multiplicities for Representations of Simple Lie Algebras (1985)
91. *A. E. Fekete*, Real Linear Algebra (1985)
92. *S. B. Chae*, Holomorphy and Calculus in Normed Spaces (1985)
93. *A. J. Jerri*, Introduction to Integral Equations with Applications (1985)
94. *G. Karpilovsky*, Projective Representations of Finite Groups (1985)
95. *L. Narici and E. Beckenstein*, Topological Vector Spaces (1985)
96. *J. Weeks*, The Shape of Space (1985)
97. *P. R. Gribik and K. O. Kortanek*, Extremal Methods of Operations Research (1985)
98. *J.-A. Chao and W. A. Woyczynski, eds.*, Probability Theory and Harmonic Analysis (1986)
99. *G. D. Crown et al.*, Abstract Algebra (1986)
100. *J. H. Carruth et al.*, The Theory of Topological Semigroups, Volume 2 (1986)
101. *R. S. Doran and V. A. Belfi*, Characterizations of C*-Algebras (1986)
102. *M. W. Jeter*, Mathematical Programming (1986)
103. *M. Altman*, A Unified Theory of Nonlinear Operator and Evolution Equations with Applications (1986)
104. *A. Verschoren*, Relative Invariants of Sheaves (1987)

Additional Volumes in Preparation

REAL FUNCTION ALGEBRAS

S. H. Kulkarni
Indian Institute of Technology
Madras, India

B. V. Limaye
Indian Institute of Technology
Bombay, India

CRC Press
Taylor & Francis Group
Boca Raton London New York

CRC Press is an imprint of the
Taylor & Francis Group, an **informa** business

CRC Press
Taylor & Francis Group
6000 Broken Sound Parkway NW, Suite 300
Boca Raton, FL 33487-2742

First issued in paperback 2019

© 1992 by Taylor & Francis Group, LLC
CRC Press is an imprint of Taylor & Francis Group, an Informa business

No claim to original U.S. Government works

ISBN-13: 978-0-8247-8653-3 (hbk)
ISBN-13: 978-0-367-40271-6 (pbk)

Library of Congress Cataloging-in-Publication Data

Kulkarni, S. H.
        Real function algebras / S. H. Kulkarni, B. V. Limaye.
        p.   cm. — (Monographs and textbooks in pure and applied mathematics : 168)
        Includes bibliographical references and index.
        ISBN 0-8247-8653-X
        1. Function algebras.   I. Limaye, Balmohan Vishnu.   II. Title.
III. Series.
        QA326.K85   1992
        512′.55—dc20
                                                                                                92-4141
                                                                                                CIP

Visit the Taylor & Francis Web site at
http://www.taylorandfrancis.com

and the CRC Press Web site at
http://www.crcpress.com

Preface

The beautiful theory of complex function algebras has been developed extensively over the past thirty years or so. These are the uniformly closed complex subalgebras of $C(X)$ separating the points of X and containing the constants, where X is a compact Hausdorff space. The enrichment of this theory has come largely from the theory of complex analytic functions. This area now occupies a proud place in the general theory of complex Banach algebras. On the other hand, real function algebras (and, for that matter, real Banach algebras in general) have received relatively scant attention. These are the uniformly closed real subalgebras of $C(X,\tau)$ separating the points of X and containing the constants, where τ is a continuous involution on X and $C(X,\tau) = \{f \in C(X) : f \circ \tau = \bar{f}\}$.

Our aim in writing this book is to give an account of the present theory of real function algebras. This theory has been developed essentially during the last ten years in analogy with the theory of complex function algebras. There are two distinct approaches to this development. One is the so-called *complexification technique*: If A is a real function algebra on (X,τ), then its complexification $B = \{f + ig : f,g \text{ in } A\}$ is a complex function algebra on X, and results for A can often be deduced from the corresponding results for B. The other approach is to prove results for A intrinsically, without resorting to its complexification. Although the intrinsic approach is, in general, more difficult than the former, we have preferred it on aesthetic grounds. Also, since every complex function algebra can be viewed as a real function algebra, this approach at once generalizes the theory of complex function algebras. Only when the intrinsic approach seemed too difficult is recourse made to the complexification technique. A reader of this book is not required to be already conversant with the theory of complex function algebras, except in isolated instances. In fact, even the prerequisites from the theory of real and complex Banach algebras are developed in the first three sections of Chapter 1. The book is addressed to those readers who have a brief introduction to real and complex variables and to functional analysis. It may enable them to employ their knowledge to obtain further results concerning continuous, analytic, and harmonic functions.

At the beginning of each chapter, we have summarized its contents. The dependence of various sections of the book is indicated in the following diagram.

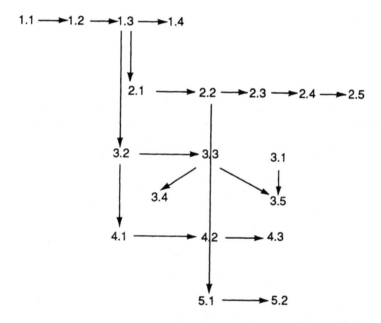

The present extent of the theory of real function algebras (and hence the scope of this book) is rather limited. Real analogs of several interesting complex results remain to be explored. Even among the known real results, we discuss only those that are in our view basic. However, we have tried to be exhaustive in giving references to related results.

It seems worthwhile to point out some results in this book that are of general mathematical interest: topological conditions for the commutativity of a real or complex Banach algebra (Theorem 1.1.30), Ransford's short elementary proof of the Bishop–Stone–Weierstrass theorem (Theorem 2.1.2), a result implying the analyticity or the antianalyticity of a function f from the harmonicity of $\operatorname{Re} f$, $\operatorname{Re} f^2$, $\operatorname{Re} f^3$, and $\operatorname{Re} f^4$ (Theorem 3.1.3), and a result regarding the positivity of a real linear functional on a real subspace of $C(X)$ (Lemma 3.2.4).

We thank the Directors of the Indian Institutes of Technology at Bombay and Madras for making it possible for us to work together under the Inter I.I.T. Faculty Exchange Programme. We acknowledge the support of the Curriculum Development Cell at the I.I.T. Bombay and the Centre for Continuing Education at the I.I.T. Madras in preparing the manuscript, and of Marcel Dekker, Inc., New York, in publishing this book. We are indebted to R. C. Burckel and K. Jarosz for their encouragement. We also thank M. Parameswaran and M. M. Thomas for typing the manuscript, and Sushama Aggrawal, S. Arundhathi, K. C. Sivakumar, M. H. Vasavada, and P. Veeramani for reading it critically. Finally, we are grateful to Chaya Kulkarni and Nirmala Limaye, who kept our morale high during the preparation of this work.

S. H. Kulkarni
B. V. Limaye

Contents

1
Introduction

In this monograph we are concerned with real function algebras, which are, in particular, Banach algebras. For this reason we begin by developing the necessary theory of real Banach algebras. Another reason for developing this general theory is that in the study of real function algebras, we have to deal with some Banach algebras that are not real function algebras.

The first section contains a definition and some examples of Banach algebras, properties of the spectrum of an element, the spectral radius formula, functional calculus, and some conditions for the commutativity of a Banach algebra. The second section deals with the representation of a commutative real Banach algebra with a unit element as an algebra of continuous functions on a compact Hausdorff space. We define a real function algebra in the third section and give several examples. Some properties of real function algebras are also discussed. These three sections are an essential prerequisite for the remainder of the book. The final section introduces the relatively recently developed topic of perturbation of Banach algebras. It is not used later in this book.

1.1 ELEMENTS OF REAL BANACH ALGEBRAS

Definition 1.1.1 *An algebra A* over a field $\mathbb{F}$ is a ring which is also a vector space over $\mathbb{F}$ such that $(\alpha a)(b) = \alpha(ab) = (a)(\alpha b)$ for all a,b in A and α in $\mathbb{F}$. A is said to be *commutative* if $ab = ba$ for a and b in A.

An algebra A is called a *real algebra* if $\mathbb{F} = \mathbb{R}$, the field of real numbers, and a *complex algebra* if $\mathbb{F} = \mathbb{C}$, the field of complex numbers; such an algebra is said to be *normed* by $\|\cdot\|$ if $(A, \|\cdot\|)$ is a normed linear space such that

$$\|ab\| \leq \|a\|\,\|b\|$$

for all a,b in A; if A has a unit element 1 (i.e., $1a = a = a1$ for all a in A), it is unique and $\|1\| \geq 1$. We can, in fact, assume that $\|1\| = 1$ since $\|a\|_1 = \sup\{\|ab\|/\|b\| : 0 \neq b \text{ in } A\}$ is an equivalent norm on A for which $\|1\|_1 = 1$.

If every Cauchy sequence in a normed algebra A converges in A, then A is called a *Banach algebra*. Note that every complex Banach algebra A can be considered as a real Banach algebra A_R. Later we shall consider a natural way of constructing a complex Banach algebra from a real Banach algebra.

Example 1.1.2 Let X be a compact Hausdorff topological space. By $C(X)$ we denote the set of all complex-valued continuous functions on X. Under pointwise operations, $C(X)$ is a complex commutative algebra and has the constant function 1 as the unit element. For $f \in C(X)$, we define

$$\|f\| := \sup\{|f(x)| : x \in X\}.$$

It is easy to see that $C(X)$ is a complex Banach algebra.

As examples of closed real subalgebras that are not complex subalgebras, we may consider the following.
(a) Let Y be a closed subset of X. Consider

$$C_Y := \{f \in C(X) : f(Y) \subset \mathbb{R}\}.$$

If Y is the empty set, $C_Y = C(X)$. For every nonempty set Y, C_Y is a real commutative Banach algebra. When $Y = X$, we get the algebra of all continuous real-valued functions on X. We denote this algebra by $C_{\mathbb{R}}(X)$.
(b) Let $\tau : X \to X$ be a homeomorphism such that $\tau^2 = \tau \circ \tau$ is the identity map on X. Let

$$C(X,\tau) := \{f \in C(X) : f(\tau(x)) = \bar{f}(x) \quad \text{for all } x \in X\}.$$

Then $C(X,\tau)$ is a real Banach algebra. It is not a complex algebra. Throughout the book we shall have frequent occasions to deal with this algebra.

Note that if τ is the identity map on X, then $C(X,\tau) = C_{\mathbb{R}}(X)$. In fact, every example of type (a) can be transformed into one of type (b) by constructing a new space from one copy of Y and two copies of $X \backslash Y$.

However, not all examples of type (b) can be put in the form of type (a). [See Exercises 8A and 8B of Goodearl (1982) for details.]

Example 1.1.3 Let $X \neq \{0\}$ be a (real or complex) normed linear space, and let $BL(X)$ denote the set of all bounded linear operators on X. Under the usual operations $BL(X)$ is a noncommutative algebra. It has a unit element, namely, the identity operator, which we denote by I. Further, for $T \in BL(X)$, we define the *operator norm* by

$$\|T\| := \sup\{\|T(x)\| : x \in X, \quad \|x\| \leq 1\}.$$

Then $(BL(X), \|\cdot\|)$ is a normed algebra. It is a Banach algebra if and only if X is a Banach space.

In particular, if $X = \mathbb{R}^n$ (respectively, $\mathbb{C}^n$), then $BL(X)$ can be identified, in the usual way, with the algebra of all $n \times n$ matrices with real (respectively, complex) entries.

Example 1.1.4 Let $\mathbf{1} = (1,0,0,0)$, $\mathbf{i} = (0,1,0,0)$, $\mathbf{j} = (0,0,1,0)$, and $\mathbf{k} = (0,0,0,1)$ denote the basis vectors of $\mathbb{R}^4$. We define

$$\mathbf{1}^2 = \mathbf{1}, \quad \mathbf{1i} = \mathbf{i} = \mathbf{i1}, \quad \mathbf{1j} = \mathbf{j} = \mathbf{j1}, \quad \mathbf{1k} = \mathbf{k} = \mathbf{k1},$$

$$\mathbf{i}^2 = \mathbf{j}^2 = \mathbf{k}^2 = -\mathbf{1}, \quad \mathbf{ij} = -\mathbf{ji} = \mathbf{k}, \quad \mathbf{jk} = -\mathbf{kj} = \mathbf{i},$$

$$\mathbf{ki} = -\mathbf{ik} = \mathbf{j},$$

and extend this product to $\mathbb{R}^4$ by linearity. Then we obtain a real noncommutative algebra, known as the *real quaternion algebra* $\mathbb{H}$. For $a = a_0\mathbf{1} + a_1\mathbf{i} + a_2\mathbf{j} + a_3\mathbf{k}$ in $\mathbb{H}$, we define $\|a\| := (a_0^2 + a_1^2 + a_2^2 + a_3^2)^{1/2}$. It is routine to check that $(\mathbb{H}, \|\cdot\|)$ is a real Banach algebra with the unit element $\mathbf{1}$. In fact, for a, b in $\mathbb{H}$, $\|ab\| = \|a\| \|b\|$.

Example 1.1.5 Let W be the set of all complex-valued functions on $[0, 2\pi]$ whose Fourier series are absolutely convergent, that is, functions of the form

$$f(t) = \sum_{n=-\infty}^{\infty} c_n \exp(int), \quad t \in [0, 2\pi],$$

where c_n is a complex number for each n and $\sum_{n=-\infty}^{\infty} |c_n|$ is finite. For such a function f, we define

$$\|f\| = \sum_{n=-\infty}^{\infty} |c_n|.$$

Under pointwise operations, W is a complex Banach algebra. This is known as the *Wiener algebra*, partly because of the following theorem of Wiener: If $f \in W$ and $f(t) \neq 0$ for all $t \in [0, 2\pi]$, then $1/f \in W$. Gelfand gave an elegant proof of this result in his fundamental paper of 1941.

The complex Banach algebras have received far greater attention than the real Banach algebras. The main reason seems to be the possibility of using the power of highly developed analytic function theory via the Gelfand transformation. Scalar multiplication in a complex algebra A is a map from $\mathbb{C} \times A$ to A. By considering the restriction of this map to $\mathbb{R} \times A$, A can be viewed as a real algebra. Hence every complex Banach algebra is also a real Banach algebra. Thus it is natural to ask what can be said about this larger class. However, in the classical literature one can find only incidental remarks on the real scalar case. There are a few exceptions to this pattern: the articles of Arens and Kaplansky (1948), Kadison (1951), and Kaplansky (1949). In Rickart's treatise (1960), a large part of the theory is developed for real and complex scalars simultaneously. The books of Bonsall and Duncan (1973) and Zelazko (1973) contain a few sections about real Banach algebras. About a third of the monograph of Goodearl (1982) is devoted to a particular type of real Banach algebras: real C*-algebras. The first systematic exposition of real Banach algebras was given by Ingelstam (1964). Ingelstam has dealt with many problems in real Banach algebras (1962, 1963, 1966, 1967, 1969). In the 1970s, further attention to real Banach algebras was paid in Alling (1970), Alling and Greenleaf (1971), Alling and Campbell (1972), Limaye (1975), Limaye and Simha (1975), Oliver (1970), Ono (1970), and Palmer (1972).

We now proceed to develop some parts of the theory of real Banach algebras by first describing a procedure for studying a real Banach algebra via its complexification.

Definition 1.1.6 Given a real algebra A, the *complexification B of A* is the set $A \times A$ with the operations of addition, multiplication, and scalar multiplication defined by

$$(a,b) + (c,d) := (a + c, b + d),$$
$$(\alpha + i\beta)(a,b) := (\alpha a - \beta b, \alpha b + \beta a),$$
$$(a,b)(c,d) := (ac - bd, ad + bc),$$

for all a, b, c, d in A and α, β in $\mathbb{R}$.

It is routine to check that B is a complex algebra. If $(A, \|\cdot\|)$ is a real normed algebra, it is possible to construct a norm p on B satisfying the following properties:

(i) (B,p) is a complex normed algebra and (B,p) is Banach whenever $(A, \|\cdot\|)$ is Banach.

(ii) $\max(\|a\|, \|b\|) \le p((a,b)) \le 2 \max(\|a\|, \|b\|)$ for a,b in A.

(iii) $p((a,0)) = \|a\|$ for a in A.

Since we shall not have any occasion to use this *normed complexification* of an arbitrary real normed algebra in this book, we do not bother to prove the foregoing assertion. Interested readers can find a proof in Bonsall and Duncan (1973). A usual method of studying a real Banach algebra A is to consider the complexification B of A and then to apply known techniques of complex Banach algebras to B in order to deduce results for the original algebra A. We shall refer to this method as the *complexification technique*. Arens and Kaplansky (1948), Rickart (1960), Bonsall and Duncan (1973), Alling (1970), and Ingelstam (1964) all use this technique primarily.

By an *intrinsic method* of studying real Banach algebras, we mean a method that does not resort to the complexification technique. It has a pedagogic advantage in the sense that one never goes outside the structure of a real Banach algebra and works only within it; however, there is no bar to using complex numbers and functions while employing the intrinsic method.

The concept of the spectrum of an element is basic in the theory of Banach algebras.

Definition 1.1.7 Let A be a real or complex algebra with unit 1 and $a \in A$. An element b (respectively, c) is called a *left* (respectively, *right*) *inverse* of a if $ba = 1$ (respectively, $ac = 1$). An *inverse* of a is both a left and a right inverse of a. If a has an inverse, a is called *invertible* or *regular*; otherwise, it is called *singular*.

Definition 1.1.8 Let A be a complex algebra with unit 1 and $a \in A$. Then the *spectrum of a in A* is the subset $Sp(a,A)$ of $\mathbb{C}$ defined as follows:

$$Sp(a,A) := \{\lambda \in \mathbb{C} : a - \lambda \text{ is singular in } A\}.$$

Remark 1.1.9 Note that we have used the same symbol λ for the scalar λ and the element $\lambda \cdot 1$ of A. We shall continue this practice as the meaning is clear from the context. Further, we shall denote $Sp(a,A)$ simply by $Sp(a)$, when no emphasis on the algebra A is required. This happens, for example, when we consider the spectra of various elements in the same algebra.

It is natural to think that for a real algebra A with unit 1, the spectrum of an element a in A should be defined as the set of all real λ for which $a - \lambda$ is singular in A. But if we accept this definition, the spectra of many elements may turn out to be the empty set. For example, if A is the algebra of all 2×2 real matrices, and if

$$a = \begin{bmatrix} 0 & 1 \\ -1 & 0 \end{bmatrix},$$

then

$$a - \lambda = \begin{bmatrix} -\lambda & 1 \\ -1 & -\lambda \end{bmatrix}$$

is invertible for every real λ. Hence we adopt the following definition due to Kaplansky (1949).

Definition 1.1.10 Let A be a real algebra with unit 1. For $a \in A$, *the spectrum of a in A* is the subset $\text{Sp}(a, A)$ of $\mathbb{C}$ defined as follows:

$$\text{Sp}(a, A) := \{s + it \in \mathbb{C} : (a - s)^2 + t^2 \text{ is singular in } A\}.$$

Clearly, $s + it \in \text{Sp}(a, A)$ if and only if $s - it \in \text{Sp}(a, A)$.

Remark 1.1.11 The following facts can be deduced in a straightforward manner from Definitions 1.1.8 and 1.1.10.
(i) If A is a complex algebra with unit and if A_R denotes A regarded as a real algebra, then

$$\text{Sp}(a, A_R) = \text{Sp}(a, A) \cup \{\bar{\lambda} : \lambda \in \text{Sp}(a, A)\}.$$

(ii) If A is a real algebra with unit and B is the complexification of A (as in Definition 1.1.6), then

$$\text{Sp}(a, A) = \text{Sp}((a, 0), B) \qquad \text{for every } a \text{ in } A.$$

Definition 1.1.12 Let A be a real normed algebra and let $a \in A$. The *spectral radius $r(a)$ of a* is defined by

$$r(a) := \inf\{\|a^n\|^{1/n} : n = 1, 2, \ldots\}.$$

The reason for the nomenclature ''spectral radius'' will be clear later. Note that $r(a) \leq \|a\|$. Also, if $ab = ba$ for a, b in A, then $r(ab) \leq r(a)r(b)$ since $(ab)^n = a^n b^n$ for all $n = 1, 2, \ldots$.

Lemma 1.1.13 Let A be a real normed algebra and let $a \in A$. Then

$$r(a) = \lim_{n \to \infty} \|a^n\|^{1/n}.$$

Proof. By definition, $r(a) \leq \|a^n\|^{1/n}$ for all n. Now let $\varepsilon > 0$. Then there exists k such that $\|a^k\|^{1/k} < r(a) + \varepsilon$. By the division algorithm, for every natural number n, there exist unique nonnegative integers $p(n)$ and $q(n)$ such that $n = p(n)k + q(n)$ and $q(n) \leq k - 1$. As $n \to \infty$, $q(n)/n \to 0$, hence $p(n)k/n \to 1$. Thus we have

$$\|a^n\|^{1/n} \leq \|a^k\|^{p(n)/n}\|a\|^{q(n)/n} \longrightarrow \|a^k\|^{1/k} < r(a) + \varepsilon.$$

This completes the proof. □

Now we show that if $ab = ba$, then $r(a + b) \leq r(a) + r(b)$. The following proof, adapted from Rickart (1960), uses Lemma 1.1.13.

Let $\alpha > r(a)$, $\beta > r(b)$, $x = a/\alpha$, and $y = b/\beta$, so that $r(x), r(y) < 1$. Since $ab = ba$, we have for $n = 1,2,\ldots,$

$$\|(a + b)^n\|^{1/n} = \left\|\sum_{k=0}^{n} {}_nC_k a^k b^{n-k}\right\|^{1/n}$$

$$\leq \left[\sum_{k=0}^{n} {}_nC_k \alpha^k \beta^{n-k}\|x^k\| \, \|y^{n-k}\|\right]^{1/n}.$$

For each n, let k_n be such that $0 \leq k_n \leq n$ and

$$\|x^{k_n}\| \, \|y^{n-k_n}\| = \max_{0 \leq k \leq n} \|x^k\| \, \|y^{n-k}\|.$$

Then it follows that for $n = 1,2,\ldots,$

$$r(a + b) \leq \|x^{k_n}\|^{1/n}\|y^{n-k_n}\|^{1/n}(\alpha + \beta).$$

Since $0 \leq k_n/n \leq 1$, we choose a subsequence (n_m) of natural numbers such that if $k'_m = k_{n_m}$, then (k'_m/n_m) converges to δ, say. Then $0 \leq \delta \leq 1$. If $\delta = 0$, then

$$\varlimsup_{m \to \infty} \|x^{k'_m}\|^{1/n_m} \leq \varlimsup_{m \to \infty} \|x\|^{k'_m/n_m} = 1,$$

and if $0 < \delta \leq 1$, then $k'_m \neq 0$ for all large m, so that

$$\varlimsup_{m \to \infty} \|x^{k'_m}\|^{1/n_m} = \varlimsup_{m \to \infty} (\|x^{k'_m}\|^{1/k'_m})^{k'_m/n_m} = r(x)^\delta \leq 1.$$

Similarly, $\varlimsup_{m \to \infty} \|y^{k'_m}\|^{1/n_m} \leq 1$. Hence $r(a + b) \leq \alpha + \beta$. Since this is true for all $\alpha > r(a)$ and $\beta > r(b)$, we have $r(a + b) \leq r(a) + r(b)$.

Lemma 1.1.14 Let A be a real Banach algebra with unit 1, let $a \in A$, and let s and t be real numbers.
(i) If $r(a) < |s|$, then $a - s$ is regular in A and

$$(a - s)^{-1} = -\sum_{n=0}^{\infty} \frac{a^n}{s^{n+1}}.$$

(ii) If $r(a) < (s^2 + t^2)^{1/2} := q$, then $(a - s)^2 + t^2$ is regular in A and its inverse is given by the absolutely convergent series $\sum_{n=0}^{\infty} b_n$, where

$$b_n = \begin{cases} \dfrac{(n + 1)a^n}{s^{n+2}} & \text{if } t = 0, \\[3mm] \dfrac{[\cos n\theta - \cos(n + 2)\theta]a^n}{(1 - \cos 2\theta)q^{n+2}} & \text{if } t \neq 0, \ \theta = \tan^{-1}\dfrac{t}{s}, \ 0 \leq \theta < \pi. \end{cases}$$

Proof.

(i) Let $r(a) < |s|$. There exists p such that $r(a) < p < |s|$. By Lemma 1.1.13, $\|a^n\| \leq p^n$ for all sufficiently large n. For such n, we have

$$\left\| \frac{a^n}{s^{n+1}} \right\| \leq \frac{p^n}{|s|^{n+1}},$$

so that $\sum_{n=0}^{\infty} a^n/s^{n+1}$ converges absolutely. Since A is Banach, let

$$c = -\sum_{n=0}^{\infty} \frac{a^n}{s^{n+1}} \quad \text{and} \quad c_m = -\sum_{n=0}^{m} \frac{a^n}{s^{n+1}}.$$

Then

$$c_m(a - s) = -\left(\frac{1}{s} + \frac{a}{s^2} + \cdots + \frac{a^m}{s^{m+1}} \right)(a - s)$$

$$= 1 - \frac{a^{m+1}}{s^{m+1}} = (a - s)c_m.$$

Since $(a/s)^m$ tends to zero as m tends to infinity, it follows that

$$c(a - s) = 1 = (a - s)c,$$

that is, c is the inverse of $a - s$.

(ii) Let $r(a) < q = (s^2 + t^2)^{1/2}$. If $t = 0$, the conclusion follows from (i). Now let $t \neq 0$, so that $1 - \cos 2\theta \neq 0$. As above, we can find p such that $r(a) < p < q$ and $\|a^n\| \leq p^n$ for all large n. For such n,

$$\|b_n\| \leq (q|\sin \theta|)^{-2} \left(\frac{p}{q} \right)^n.$$

Hence $\sum_{n=0}^{\infty} b_n$ converges absolutely. Let $d = \sum_{n=0}^{\infty} b_n$ and $d_m = \sum_{n=0}^{m} b_n$. Then a tedious but routine computation shows that

$$d_m(a^2 - 2qa \cos \theta + q^2) = 1 - \frac{[\cos(m + 1)\theta - \cos(m + 3)\theta]a^{m+1}}{(1 - \cos 2\theta)q^{m+1}}$$

$$+ \frac{[\cos m\theta - \cos(m + 2)\theta]a^{m+2}}{(1 - \cos 2\theta)q^{m+2}}$$

$$= (a^2 - 2qa \cos \theta + q^2)d_m.$$

Since $(a/q)^m$ tends to zero as m tends to infinity, we get

$$d[(a - s)^2 + t^2] = 1 = [(a - s)^2 + t^2]d;$$

that is, d is the inverse of $(a - s)^2 + t^2$. $\qquad\qquad\qquad\qquad\qquad\square$

Corollary 1.1.15 Let A be a complex Banach algebra with unit 1. Let $a \in A$ and $\lambda \in \mathbb{C}$. If $r(a) < |\lambda|$, then $a - \lambda$ is regular in A and

$$(a - \lambda)^{-1} = -\sum_{n=0}^{\infty} \frac{a^n}{\lambda^{n+1}}.$$

Proof. Let $\lambda = s + it$. If $t = 0$, the conclusion follows from the first part of Lemma 1.1.14. Now let $t \neq 0$. Note that $(a - s)^2 + t^2 = (a - \lambda)(a - \bar{\lambda})$ is invertible and its inverse is $d = \sum_{n=0}^{\infty} b_n$, as given in the second part of the lemma. Thus $(a - \lambda)(a - \bar{\lambda})d = 1 = (a - \bar{\lambda})d(a - \lambda)$. This implies that $(a - \lambda)$ is invertible and its inverse is

$$(a - \bar{\lambda})d = \sum_{n=0}^{\infty} (a - s)b_n + i \sum_{n=0}^{\infty} tb_n.$$

A simple calculation shows that the coefficient of a^n in $\sum_{n=0}^{\infty} (a - s)b_n$ is $-[\cos(n + 1)\theta]/q^{n+1}$ and the coefficient of a^n in $\sum_{n=0}^{\infty} tb_n$ is $[\sin(n + 1)\theta]/q^{n+1}$, where $q = (s^2 + t^2)^{1/2} = |\lambda|$. hence

$$(a - \lambda)^{-1} = (a - \bar{\lambda})d = -\sum_{n=0}^{\infty} \frac{a^n}{\lambda^{n+1}}. \qquad\qquad\square$$

Corollary 1.1.16 Let A be a real Banach algebra with unit 1. Suppose that $a \in A$ is invertible and $b \in A$ is such that $\|b - a\| \leq \varepsilon/\|a^{-1}\|$ with $0 \leq \varepsilon < 1$. Then b is invertible and $\|b^{-1} - a^{-1}\| \leq \|a^{-1}\|^2\|b - a\|/(1 - \varepsilon)$.

Proof.

$$r(1 - a^{-1}b) \leq \|1 - a^{-1}b\| = \|a^{-1}a - a^{-1}b\|$$

$$\leq \|a^{-1}\| \|a - b\| \leq \varepsilon < 1.$$

Hence, by (i) of Lemma 1.1.14, $-(1 - a^{-1}b - 1) = a^{-1}b$ is regular and we can easily conclude that b is also regular. Further,

$$\|b^{-1}\| - \|a^{-1}\| \le \|b^{-1} - a^{-1}\| = \|b^{-1}(a - b)a^{-1}\| \le \varepsilon\|b^{-1}\|.$$

Hence $\|b^{-1}\| \le \|a^{-1}\|/(1 - \varepsilon)$. Thus we have

$$\|b^{-1} - a^{-1}\| \le \|b^{-1}\|\,\|a - b\|\,\|a^{-1}\|$$
$$\le \frac{\|a^{-1}\|^2\|a - b\|}{1 - \varepsilon}. \qquad \square$$

Remark 1.1.17 Let A be a real Banach algebra with unit 1. Let $\mathrm{Inv}(A)$ denote the set of all invertible elements in A. Thus $\mathrm{Inv}(A)$ is a group under multiplication. Corollary 1.1.16 says that $\mathrm{Inv}(A)$ is an open set in A and the map $a \mapsto a^{-1}$ is continuous on $\mathrm{Inv}(A)$. In short, $\mathrm{Inv}(A)$ is a topological group.

Corollary 1.1.18 Let A be a real Banach algebra with unit 1 and $a \in A$. Then $\mathrm{Sp}(a, A)$ is a compact subset of $\mathbb{C}$.

Proof. By Lemma 1.1.14, if $s + it \in \mathrm{Sp}(a)$, then $(s^2 + t^2)^{1/2} \le r(a) \le \|a\|$. Hence $\mathrm{Sp}(a)$ is bounded. Thus it is enough to show that $\mathrm{Sp}(a)$ is closed. We show that its complement in $\mathbb{C}$ is open. Let $s + it \in \mathbb{C}\setminus\mathrm{Sp}(a)$. Then $(a - s)^2 + t^2 \in \mathrm{Inv}(A)$. The map $f: \mathbb{C} \to A$ defined by $f(x + iy) = (a - x)^2 + y^2$ is continuous and $\mathrm{Inv}(A)$ is an open neighborhood of $f(s + it)$. Hence there exists an open neighborhood U of $s + it$ such that $f(U) \subset \mathrm{Inv}(A)$. But then $U \subset \mathbb{C}\setminus\mathrm{Sp}(a)$. Thus $\mathbb{C}\setminus\mathrm{Sp}(a)$ is open. $\qquad \square$

Theorem 1.1.19 (Spectral radius formula) Let A be a real Banach algebra with unit 1 and $a \in A$. Then

$$r(a) = \sup\{(s^2 + t^2)^{1/2} : s + it \in \mathrm{Sp}(a, A)\}.$$

In particular, $\mathrm{Sp}(a, A)$ is nonempty.

Proof. Let

$$\alpha := \sup\{(s^2 + t^2)^{1/2} : s + it \in \mathrm{Sp}(a, A)\}.$$

Lemma 1.1.14 shows that $r(a) \ge \alpha$. We now prove that $r(a) \le \alpha$. First, let $r(a) = 0$. If $0 \notin \mathrm{Sp}(a, A)$, then a is invertible. Since a and a^{-1} commute, we have $1 = r(1) = r(aa^{-1}) \le r(a)r(a^{-1}) = 0$, which is absurd. Therefore, $0 \in \mathrm{Sp}(a, A)$, so that $\alpha \ge 0 = r(a)$.

Now, let $r(a) > 0$ and ϕ be a continuous linear functional on A. For $s + it$ in $\mathbb{C}\setminus\mathrm{Sp}(a, A)$, we define

$$u(s, t) = \phi[(a - s)((a - s)^2 + t^2)^{-1}]$$

and

$$v(s,t) = \phi[t((a - s)^2 + t^2)^{-1}].$$

It can be verified that u and v both satisfy Laplace's equation and are, in fact, harmonic conjugates of each other in $\mathbb{C} \backslash \text{Sp}(a, A)$. For $s + it$ in $\mathbb{C}$ with $r(a) < (s^2 + t^2)^{1/2}$, we have

$$u(s,t) = \sum_{n=0}^{\infty} \phi[(a - s)b_n]$$

and

$$v(s,t) = \sum_{n=0}^{\infty} \phi(tb_n),$$

where b_n, $n = 0,1,2,...$, are as in Lemma 1.1.14. Further, we have already noted in Corollary 1.1.15 that the coefficient of a^n in $(a - s)\Sigma_{n=0}^{\infty} b_n$ is $-\cos(n + 1)\theta/q^{n+1}$ and in $\Sigma_{n=0}^{\infty} tb_n$ is $\sin(n + 1)\theta a^n/q^{n+1}$, where $q = (s^2 + t^2)^{1/2}$ and $t/s = \tan \theta$, $0 \le \theta < \pi$. If we let $w = u + iv$ and $z = s + it$, then

$$w(z) = -\sum_{n=0}^{\infty} \frac{\phi(a^n)}{z^{n+1}},$$

where the series converges absolutely for $|z| > r(a)$.

Since w is an analytic function in $\mathbb{C} \backslash \text{Sp}(a, A)$, Laurent's theorem shows that the series above converges for every z with $|z| > \alpha$. For every such z, $\phi((a/|z|)^n)$ is bounded. Since this is true for every continuous linear functional ϕ on A, by the uniform boundedness principle, there exists M such that $\|(a/|z|)^n\| \le M$. Hence for every z with $|z| > \alpha$,

$$r(a) = \lim_{n \to \infty} \|a^n\|^{1/n} \le \lim_{n \to \infty} M^{1/n}|z| = |z|.$$

Thus $r(a) \le \alpha$. $\qquad\qquad\qquad\qquad\qquad\qquad\qquad\qquad\qquad \square$

Remark 1.1.20 The proof above was given in Kulkarni and Limaye (1980). A proof using the complexification technique can be found in Ingelstam (1964), Rickart (1960), and Bonsall and Duncan (1973).

In view of Remark 1.1.11(i), the spectral radius formula is also valid for complex Banach algebras. Note that the spectral radius formula equates an algebraically obtained number $\sup\{(s^2 + t^2)^{1/2} : s + it \in \text{Sp}(a, A)\}$ with $r(a)$, which depends on the norm. This is a good example of the interplay between the algebraic and topological structures of A.

We now give some applications of the spectral radius formula. Let A and B be algebras over a field $\mathbb{F}$. Recall that a linear map $T : A \to B$ is called a *homomorphism* if $T(ab) = T(a)T(b)$ for all a, b in A. A homomorphism is an *isomorphism* if it is $1 : 1$ and onto. A is said to be *isomorphic* to B if there exists an isomorphism from A onto B. Let X and Y be normed linear spaces over $\mathbb{R}$ or $\mathbb{C}$. Then a linear map $S : X \to Y$ is called an *isometry* if $\|S(x)\| = \|x\|$ for every x in X.

An *isometric isomorphism* between two normed algebras is a map that is both an isometry and an isomorphism. Whenever such a map exists between two normed algebras, they are called *isometrically isomorphic* to each other.

Corollary 1.1.21 Let A and B be real Banach algebras with units and $T : A \to B$ be a homomorphism. Then $r(T(a)) \le r(a)$ for every a in A.

Proof. We denote the units of both A and B by the same symbol 1. It can easily be proved that $T(1) = 1$. Now let $s + it \notin \mathrm{Sp}(a, A)$. Then there is b in A such that

$$[(a - s)^2 + t^2]b = 1 = b[(a - s)^2 + t^2].$$

Hence

$$[(T(a) - s)^2 + t^2]T(b) = 1 = T(b)[(T(a) - s)^2 + t^2].$$

Thus $\mathrm{Sp}(T(a), B) \subset \mathrm{Sp}(a, A)$, and the result follows from 1.1.19. $\square$

Remark 1.1.22
(i) If T is an isomorphism, then $r(T(a)) = r(a)$ for every a in A.
(ii) Suppose that B has the property $\|b\| \le kr(b)$ for every b in B. A normed algebra satisfying this property is called a *spectrally normed algebra*. Algebras in Example 1.1.2 and 1.1.4 are spectrally normed. Then

$$\|T(a)\| \le kr(T(a)) \le kr(a) \le k\|a\|, \qquad a \in A.$$

Thus every homomorphism of A into a spectrally normed Banach algebra is continuous. In particular, every homomorphism of A into $\mathbb{R}$ or $\mathbb{C}$ is continuous. We now prove a fundamental theorem that characterizes normed division algebras. Recall that an algebra with unit is called a *division algebra* if every nonzero element is regular.

Theorem 1.1.23 (Mazur–Gelfand theorem) Let A be a Banach division algebra.
(i) If A is a complex algebra, then A is isometrically isomorphic to $\mathbb{C}$.

(ii) If A is a real algebra such that

$$x, y \in A \text{ and } x^2 + y^2 = 0 \quad \text{imply that} \quad x = y = 0, \qquad (*)$$

then A is isometrically isomorphic to $\mathbb{R}$.

(iii) If A is a commutative real algebra and A does not satisfy (*), then A is isomorphic to $\mathbb{C}$.

Proof. In view of Theorem 1.1.19, $\text{Sp}(a, A)$ is nonempty for every a in A.

(i) Let $a \in A$ and $\lambda \in \text{Sp}(a, A)$. Then $a - \lambda$ is singular. Hence $a - \lambda = 0$. Thus $a = \lambda \cdot 1$ and $\|a\| = |\lambda| \|1\| = |\lambda|$.

(ii) Let $a \in A$ and $s + it \in \text{Sp}(a, A)$. Then $(a - s)^2 + t^2$ is singular. Hence $(a - s)^2 + t^2 = 0$. By (*) this implies that $t = 0$ and $a = s$. Also, $\|a\| = |s|$. This proves (ii).

(iii) Since A does not satisfy (*), there exist x, y in A such that $x^2 + y^2 = 0$ and at least one of (and hence both) x, y are nonzero. Let $j = xy^{-1}$. Then by commutativity $1 + j^2 = 0$. Now let $a \in A$ and $s + it \in \text{Sp}(a, A)$. Then, as above, $(a - s)^2 + t^2 = 0$, so that

$$(a - (s + tj))(a - (s - tj)) = (a - s)^2 + t^2 = 0.$$

Hence $a = s + tj$ or $a = s - tj$, as A is a division algebra. Thus $A = \text{Span}\{1, j\}$ and $j^2 = -1$, that is, A is isomorphic to $\mathbb{C}$. $\quad\square$

We remark that if A is a real noncommutative Banach division algebra, then A is isomorphic to $\mathbb{H}$ (Example 1.1.4): Let $a \in A$ and $s + it \in \text{Sp}(a, A)$. Then $(a - s)^2 + t^2$ is singular and hence is 0. This shows that every element of A is a root of a quadratic polynomial with real coefficients. Hence A is finite dimensional by a theorem of Wedderburn. Again, by a theorem of Frobenius, every finite-dimensional real division algebra is isomorphic to $\mathbb{R}$, $\mathbb{C}$, or $\mathbb{H}$. Since A is noncommutative, A is isomorphic to $\mathbb{H}$. Ingelstam (1964) and Bonsall and Duncan (1973) give a proof of this fact without making explicit use of the theorems of Wedderburn and Frobenius.

Let A be a real Banach algebra and $a \in A$. Let f be a complex-valued function defined on some subset of the complex plane. Under certain conditions on f, we can define an element in A, which can justifiably be denoted by $f(a)$. A study of such conditions and of the relationship between a and $f(a)$ is called *functional calculus*. We begin this study with polynomials, the simplest functions that one can think of.

Definition 1.1.24 A subset D of $\mathbb{C}$ is said to be *symmetric about* the *real axis* if $\bar{z} \in D$ for every z in D.

Let $D \subset \mathbb{C}$ be symmetric about the real axis. By $P_R(D)$ we denote the algebra of all polynomials on D with real coefficients. We may note that if D is an infinite set, then a polynomial p with complex coefficients belongs to $P_R(D)$ if and only if $p(\bar{z}) = \bar{p}(z)$ for every z in D.

Let A be a real Banach algebra with unit and let $a \in A$. We have already noted that $\mathrm{Sp}(a, A)$ is symmetric about the real axis. Given p in $P_R(D)$ with $p(z) = \alpha_0 + \alpha_1 z + \cdots + \alpha_n z^n$, z in D, we denote by $p(a)$ the element of A given by

$$p(a) = \alpha_0 + \alpha_1 a + \cdots + \alpha_n a^n.$$

Theorem 1.1.25 Let A be a real Banach algebra with unit 1, $a \in A$, and D be an open neighborhood of $\mathrm{Sp}(a, A)$, which is symmetric about the real axis. Then the mapping $p \mapsto p(a)$ is a homomorphism of $P_R(D)$ into A which satisfies

$$\mathrm{Sp}(p(a), A) = \{p(z) : z \text{ in } \mathrm{Sp}(a, A)\}.$$

Proof. That $p \mapsto p(a)$ is a homomorphism of $P_R(D)$ into A is evident. Now let $\lambda \in \mathbb{C}$, and p in $P_R(D)$ be of degree n. Then there exist $\beta, \alpha_1, \ldots, \alpha_n$ in $\mathbb{C}$ such that

$$\lambda - p(z) = \beta(\alpha_1 - z) \cdots (\alpha_n - z) \qquad \text{for all } z \text{ in } D \qquad (1)$$

Since $z \in D$ implies that $\bar{z} \in D$, replacing z by $\bar{z}$ in (1)

$$\lambda - p(\bar{z}) = \beta(\alpha_1 - \bar{z}) \cdots (\alpha_n - \bar{z}) \qquad \text{for all } z \text{ in } D \qquad (2)$$

Noting that $p(\bar{z}) = \overline{p(z)}$ and taking complex conjugates of both sides in (2), we get

$$\bar{\lambda} - p(z) = \bar{\beta}(\bar{\alpha}_1 - z) \cdots (\bar{\alpha}_n - z). \qquad (3)$$

From (1) and (3), we obtain

$$(\lambda - p(z))(\bar{\lambda} - p(z)) = |\beta|^2(\alpha_1 - z)(\bar{\alpha}_1 - z) \cdots (\alpha_n - z)(\bar{\alpha}_n - z). \quad (4)$$

Let $\lambda = s + it$ and $\alpha_k = s_k + it_k$, $k = 1, 2, \ldots, n$. Then equation (4) becomes

$$(s - p(z))^2 + t^2 = |\beta|^2[(s_1 - z)^2 + t_1^2] \cdots [(s_n - z)^2 + t_n^2].$$

Since $p \mapsto p(a)$ is a homomorphism from $P_R(D)$ into A, we have

$$(s - p(a))^2 + t^2 = |\beta|^2[(s_1 - a)^2 + t_1^2] \cdots [(s_n - a)^2 + t_n^2].$$

When p is nonconstant, $\beta \neq 0$. Hence $(s - p(a))^2 + t^2$ is singular if and only if $(s_k - a)^2 + t_k^2$ is singular for some k. Thus $\lambda \in \mathrm{Sp}(p(a))$ if and only if $\lambda = p(z)$ for some z in $\mathrm{Sp}(a)$. $\qquad \square$

Let A, a, and D be as above. Note that if $p \in P_R(D)$ and $p(z) \neq 0$ for all z in D, then by Theorem 1.1.25, $p(a)$ is invertible in A. By $R_R(D)$, we denote the algebra of all rational functions p/q, where $p, q \in P_R(D)$ and $q(z) \neq 0$ for all z in D. For $f = p/q$ in $R_R(D)$ by $f(a)$ we denote the element $p(a)q(a)^{-1}$. Then in a similar way we can prove that the mapping $f \mapsto f(a)$ is a homomorphism of $R_R(D)$ into A, and it satisfies

$$\mathrm{Sp}(f(a), A) = \{f(z) : z \in \mathrm{Sp}(a, A)\}.$$

We shall now extend the considerations above to a class of functions which is bigger than that of rational functions. A detailed treatment of this topic can be found in Kulkarni and Limaye (1981a). Here we consider only some special cases.

Suppose that A is a real Banach algebra with unit 1, $a \in A$, and $\mathrm{Sp}(a) \subset D$, where D is a disk that is symmetric about the real axis. In particular, the center of D is a real number. Let α be the center of D, let ρ be the radius of D, and $H_R(D)$ denote the algebra of all analytic functions f in D satisfying $f(\bar{z}) = \bar{f}(z)$ for all z in D. Such an f can be expanded in a Taylor series around α, so that $f(z) = \Sigma_{n=0}^{\infty} \lambda_n(z - \alpha)^n$ for all z in D with λ_n in R and $\overline{\lim}_{n \to \infty} |\lambda_n|^{1/n} \leq 1/\rho$. Since $\mathrm{Sp}(a) \subset D$, we have $\mathrm{Sp}(a - \alpha) \subset \{z : |z| < \rho\}$, so that $r(a - \alpha) < \eta < \rho$. Hence for large n, $\|\lambda_n(a - \alpha)^n\| \leq (\eta/\rho)^n$. Since $\Sigma(\eta/\rho)^n$ converges, $\Sigma_{n=0}^{\infty} \lambda_n(a - \alpha)^n$ converges (absolutely) to an element of A, which we denote by $f(a)$.

Theorem 1.1.26 (Spectral mapping theorem) Let A be a real Banach algebra with unit 1, $a \in A$, and suppose that $\mathrm{Sp}(a) \subset D := \{z \in \mathbb{C} : |z - \alpha| < \rho\}$, α in R. Then the mapping $f \mapsto f(a)$ is a homomorphism of $H_R(D)$ into A, satisfying

$$\mathrm{Sp}(f(a), A) = \{f(z) : z \in \mathrm{Sp}(a, A)\}.$$

Proof. That the mapping $f \mapsto f(a)$ is a homomorphism is obvious. Now let $f \in H_R(D)$,

$$f(z) = \sum_{n=0}^{\infty} \lambda_n(z - \alpha)^n, \qquad p_m(z) := \sum_{n=0}^{\infty} \lambda_n(z - \alpha)^n, \qquad m = 0, 1, 2, \ldots.$$

λ_n in $\mathbb{R}$, z in D. Then $p_m \in P_R(D)$ for all $m = 0, 1, 2, \ldots$, $\|p_m(a) - f(a)\| \to 0$, and p_m converges uniformly to f over each closed disk contained in D.

Let $\lambda \in \mathrm{Sp}(a)$, $f(\lambda) = s + it$, and $p_m(\lambda) = s_m + it_m$. Since $p_m(a) \to f(a)$ and $s_m + it_m \to s + it$, we have $(s_m - p_m(a))^2 + t_m^2 \to (s - f(a))^2 + t^2$ as $m \to \infty$. If $s + it \notin \mathrm{Sp}(f(a))$, then $(s - f(a))^2 + t^2$ is invertible and hence $(s_m - p_m(a))^2 + t_m^2$ is invertible for large m, by Remark 1.1.17.

Hence for such m, $p_m(\lambda) = s_m + it_m \notin \mathrm{Sp}(p_m(a))$. This contradicts Theorem 1.1.25. Hence $\{f(z): z \in \mathrm{Sp}(a)\} \subset \mathrm{Sp}(f(a))$.

Next, let $z \in \mathrm{Sp}(f(a))$. Then $f(\mathrm{Sp}(a)) := \{f(\lambda): \lambda \in \mathrm{Sp}(a)\}$ is a compact set by Corollary 1.1.18. If $z \notin f(\mathrm{Sp}(a))$, then

$$\delta := \inf\{|z - f(\lambda)| : \lambda \in \mathrm{Sp}(a)\} > 0.$$

Again $\mathrm{Sp}(a)$, being compact, is contained in a closed disk contained in D, and $p_n \to f$ uniformly over each such disk. Hence we can find m_0 such that for all $m \geq m_0$ and $\lambda \in \mathrm{Sp}(a)$, we have $|f(\lambda) - p_m(\lambda)| < \delta/2$. But then, for all λ in $\mathrm{Sp}(a)$ and $m \geq m_0$,

$$|z - p_m(\lambda)| \geq |z - f(\lambda)| - |f(\lambda) - p_m(\lambda)| \geq \frac{\delta}{2} > 0.$$

In view of Theorem 1.1.25, this implies that $z \notin \mathrm{Sp}(p_m(a))$ for all $m \geq m_0$. Let $z = s + it$ and $b_m = (s - p_m(a))^2 + t^2$. Then b_m is invertible for all $m \geq m_0$. Further, by the comment following Theorem 1.1.25, we have for $m \geq m_0$,

$$\begin{aligned}
r(b_m^{-1}) &= \sup\{|((s - p_m(\lambda))^2 + t^2)^{-1}| : \lambda \in \mathrm{Sp}(a)\} \\
&= \sup\{|z - p_m(\lambda)|^{-1} |\bar{z} - p_m(\lambda)|^{-1} : \lambda \in \mathrm{Sp}(a)\} \\
&\leq \frac{4}{\delta^2}.
\end{aligned}$$

Now, let $b = (s - f(a))^2 + t^2$. Then $b_m \to b$ as $m \to \infty$. Hence we can find $m \geq m_0$ such that $\|b_m - b\| < \delta^2/4$. Note that b_m commutes with b. Hence the comment after Definition 1.1.12 shows that

$$\begin{aligned}
r(1 - b_m^{-1}b) = r(b_m^{-1}(b_m - b)) &\leq r(b_m^{-1})r(b_m - b) \\
&\leq \frac{4}{\delta^2}\|b_m - b\| < 1.
\end{aligned}$$

By Lemma 1.1.14, $b_m^{-1}b$ is invertible and hence b is invertible. This contradiction implies that

$$\mathrm{Sp}(f(a)) \subset f(\mathrm{Sp}(a)). \qquad \square$$

Remark 1.1.27 Let A be a real Banach algebra with unit 1. Since $\exp(z)$ is an entire function and $\Sigma_{n=0}^{\infty} z^n/n!$ converges uniformly to $\exp(z)$ over every closed disk with center at 0, we have $\exp(a) = \Sigma_{n=0}^{\infty} a^n/n!$. One can show that if a and b commute, then $\exp(a + b) = \exp(a)\exp(b)$. Hence $\exp(a)$ is invertible and its inverse is $\exp(-a)$. Now let

$$\begin{aligned}
\mathbb{C}^+ &:= \{s + it \in \mathbb{C} : s > 0 \text{ or } t \neq 0\} \\
&= \mathbb{C} \setminus \{t \in \mathbb{R} : t \leq 0\}.
\end{aligned}$$

For z in $\mathbb{C}^+$, $z = r\exp(i\theta)$, $r > 0$, $-\pi < \theta < \pi$. We denote by $\mathrm{Ln}(z)$ the principal branch of the logarithm of z, that is, $\mathrm{Ln}(z) = \ln r + i\theta$, $-\pi < \theta < \pi$. Then

$$\mathrm{Ln}(\exp(z)) = z \qquad \text{for all } z \text{ in } \mathbb{C}$$

and

$$\exp(\mathrm{Ln}(z)) = z \qquad \text{for all } z \text{ in } \mathbb{C}^+.$$

Thus, by Theorem 1.1.26, if $\mathrm{Sp}(a) \subset \{z \in \mathbb{C} : |z - \alpha| < \rho\} \subset \mathbb{C}^+$ for some α in $\mathbb{R}$, then $\mathrm{Ln}(a) \in A$ and

$$\exp(\mathrm{Ln}(a)) = a, \qquad \mathrm{Ln}(\exp(a)) = a.$$

Corollary 1.1.28 Let A be a real Banach algebra with unit 1 and let $a \in A$. If $r(1 - a) < 1$, then

$$\mathrm{Ln}(a) = -\sum_{n=1}^{\infty} \frac{(1 - a)^n}{n} \in A.$$

Proof. Since $r(1 - a) < 1$, $\mathrm{Sp}(a) \subset \{z \in \mathbb{C} : |z - 1| < 1\} \subset \mathbb{C}^+$. Hence $\mathrm{Ln}(a) \in A$. Also, the Taylor series expansion of $\mathrm{Ln}(z)$ around 1 is $\mathrm{Ln}(z) = -\sum_{n=1}^{\infty}(1 - z)^n/n$. Hence the result. □

We shall now give a few criteria for the commutativity of a Banach algebra. These will be used in Section 1.4. The following lemma is crucial in the proof of these criteria.

Lemma 1.1.29 Let A be a real or complex Banach algebra with unit 1. Then:
(i) $\mathrm{Sp}(ab)\setminus\{0\} = \mathrm{Sp}(ba)\setminus\{0\}$ for all a,b in A, so that $r(ab) = r(ba)$.
(ii) If there exists $k > 0$ such that $\|a\|^2 \le k\|a^2\|$ for all a in A, then A is a spectrally normed algebra, in fact; $\|a\| \le kr(a)$ for all a in A, $r(ab) = r(ba)$, and $\|ba\| \le k\|ab\|$ for all a,b in A.

Proof.
(i) First, let A be a real algebra. Let $s + it$ be in $\mathbb{C}$, $s^2 + t^2 = 1$, and $s + it$ be not in $\mathrm{Sp}(ab)$. Let c denote the inverse of $(ab - s)^2 + t^2$. Then it follows by direct calculation that $1 + b(2s - ab)ca$ is the inverse of $(ba - s)^2 + t^2$, so that $s + it \notin \mathrm{Sp}(ba)$. If A is a complex algebra, one can give a similar proof, or use Remark 1.1.11(i). The spectral radius formula (Theorem 1.1.19) shows that $r(ab) = r(ba)$.
(ii) Since $\|a\|^2 \le k\|a^2\|$ for all a in A, we have by induction

$$\|a\|^{2^n} \le k^{2^n-1}\|a^{2^n}\| \qquad \text{for all } n = 1,2,\ldots.$$

Hence $\|a\| \le kr(a)$ by taking 2^nth roots and applying Lemma 1.1.13. Now it follows from (i) that

$$\|ba\| \le kr(ba) = kr(ab) \le k\|ab\|. \qquad \square$$

Theorem 1.1.30   Suppose that A is a Banach algebra with unit 1 satisfying one of the following conditions:
(i) (Le Page, 1967; Hirschfeld and Zelazko, 1968) A is a complex algebra and there exists a positive constant k such that $\|a\|^2 \le k\|a^2\|$ for all a in A.
(ii) (Kulkarni, 1988; Srivastav, 1990) A is a real algebra and there exists a positive constant k such that $\|a\|^2 \le k\|a^2 + b^2\|$ for all a,b in A with $ab = ba$.

Then A is commutative.

Proof.   In both cases $\|a\|^2 \le k\|a^2\|$ for all a in A. Hence by (ii) of Lemma 1.1.29,

$$\|ba\| \le k\|ab\| \qquad \text{for all } a,b \text{ in } A. \tag{1}$$

Now let ϕ be a continuous linear functional on A, and let a and b be elements of A.

Suppose that A satisfies condition (i). Define $f : \mathbb{C} \to \mathbb{C}$ by

$$f(z) := \phi[\exp(za)b \exp(-za)]$$

for z in $\mathbb{C}$. Then f is an entire function. Further, by (1),

$$|f(z)| \le \|\phi\| \|\exp(za)b \exp(-za)\|$$
$$\le k\|\phi\| \|b \exp(-za) \exp(za)\| = k\|\phi\| \|b\|.$$

Thus f is bounded and hence it is a constant by Liouville's theorem. Therefore, $f(0) = f(1)$ implies that

$$\phi(b) = \phi[\exp(a)b \exp(-a)]. \tag{2}$$

Next, suppose that A satisfies condition (ii). Define $u : \mathbb{R}^2 \to \mathbb{R}^1$ by

$$u(x,y) = \phi\{\exp(xa)[(\cos ya)b \cos ya + (\sin ya)b \sin ya] \exp(-xa)\}$$

for x,y in $\mathbb{R}^2$.

A straightforward computation of partial derivatives shows that u is a harmonic function on $\mathbb{R}^2$. (The details of this computation can be found in the references cited above.) Also, for all x,y in $\mathbb{R}^2$, we have by (1),

$|u(x,y)| \leq \|\phi\| \|\exp(xa)[(\cos ya)b \cos ya + (\sin ya)b \sin ya] \exp(-xa)\|$

$\leq k\|\phi\| \|(\cos ya)b \cos ya + (\sin ya)b \sin ya\|$

$\leq k\|\phi\| \|b\|(\|\cos ya\|^2 + \|\sin ya\|^2)$

$\leq 2k^2\|\phi\| \|b\| \|\cos^2 ya + \sin^2 ya\|$

$= 2k^2\|\phi\| \|b\|.$

Thus u is a bounded harmonic function on $\mathbb{R}^2$. Hence u is a constant. Therefore, $u(0,0) = u(1,0)$ again implies equation (2).

Since equation (2) holds for every continuous linear functional ϕ on A, we have

$$b = \exp(a)b \exp(-a),$$

that is,

$$b \exp(a) = \exp(a)b \quad \text{for all } a,b \text{ in } A. \tag{3}$$

Now let c and d be any two elements of A. If $\|c\| < 1$, there exists a in A such that $1 + c = \exp(a)$ (Corollary 1.1.28). From (3) it follows that $d \exp(a) = \exp(a)d$; that is, $d(1 + c) = (1 + c)d$, or $cd = dc$. The restriction $\|c\| < 1$ can now be removed by considering $c/2\|c\|$ if $c \neq 0$. $\square$

Example 1.1.4 shows that if unlike in part (i) of Theorem 1.1.30, A is a real Banach algebra satisfying $\|a\|^2 \leq k\|a^2\|$ for all a in A and some positive content k, then A need not be commutative.

1.2 COMMUTATIVE REAL BANACH ALGEBRAS

We now proceed to develop the Gelfand theory for real commutative Banach algebras. The aim of this theory is to show that every real commutative Banach algebra with unit can be represented as an algebra of continuous functions on a compact Hausdorff space. The main theorem in this direction (Theorem 1.2.9) is due to Ingelstam (1964).

Definition 1.2.1 Let A be a real algebra. The *carrier space of A*, denoted by Car(A), is the set of all nonzero homomorphisms from A to $\mathbb{C}$, regarded as a real algebra.

Let $\phi \in \text{Car}(A)$ and define $\bar{\phi}$ by $\bar{\phi}(a) = \overline{\phi(a)}$ for a in A. Then it is easy to see that $\bar{\phi} \in \text{Car}(A)$. Let $\tau: \text{Car}(A) \to \text{Car}(A)$ be defined by $\tau(\phi) = \bar{\phi}$, $\phi \in \text{Car}(A)$.

For a in A, the *Gelfand transform* of a is the map $\hat{a}: \text{Car}(A) \to \mathbb{C}$, given by $\hat{a}(\phi) = \phi(a)$ for ϕ in Car(A). The weakest topology on Car(A) that makes $\hat{a}$ continuous on Car(A) for all a in A is called the *Gelfand topology* on Car(A).

Note that if A has unit 1, and $\phi \in \text{Car}(A)$, then $\phi(1) = 1$, because ϕ is a nonzero homomorphism. If a is invertible, then $\phi(a^{-1})\phi(a) = 1$. Thus $\phi(a)$ is nonzero for every invertible element a.

Lemma 1.2.2 Let A be a real Banach algebra with unit 1 and $\phi \in \text{Car}(A)$. Then $\|\phi\| := \sup\{|\phi(a)| : a \in A, \quad \|a\| \leq 1\} = 1$.

Proof. By Corollary 1.1.21, $|\phi(a)| \leq \|a\|$ for all a in A. Hence $\|\phi\| \leq 1$. Since $\phi(1) = 1$, we have $\|\phi\| = 1$. $\qquad\qquad\qquad\qquad\qquad\quad\square$

The following example shows that $\text{Car}(A)$ can, in general, be empty. Consider the real quaternion algebra $\mathbb{H}$ of Example 1.1.4. Suppose that $\phi \in \text{Car}(\mathbb{H})$. Then

$$\phi(2k) = \phi(ij - ji)$$
$$= \phi(i)\phi(j) - \phi(j)\phi(i) = 0.$$

Since $2k$ is invertible, we have $1 = \phi(1) = \phi(2k)\phi((2k)^{-1}) = 0$, a contradiction. Hence $\text{Car}(\mathbb{H})$ is empty. We shall, however, show in the sequel that if A is a commutative normed algebra, then $\text{Car}(A)$ is non-empty.

Definition 1.2.3 Let A be a real or complex algebra with unit. A subspace I of A is called a *left ideal* if $a \in I$ and $x \in A$ imply that $xa \in I$. Similarly, a *right ideal* is defined. I is called a *(two-sided) ideal* if I is both a left and a right ideal. An ideal (left, right, two-sided) is called *proper* if it is different from A. When A is commutative, every left or right ideal is two-sided. A *maximal ideal* is a proper ideal M which is not contained in any other proper ideal; that is, if I is an ideal and $M \subset I \subset A$, then $M = I$ or $I = A$. By a routine application of Zorn's lemma, one can prove that every (left, right, two-sided) proper ideal is contained in a maximal (respectively, left, right, two-sided) ideal.

Lemma 1.2.4 Let I be a closed two-sided ideal in a (real or complex) normed algebra A with unit. Let A/I denote the set of all cosets of the form $I + a := \{x + a : x \in I\}$. For a,b in A and α in $\mathbb{R}$ (or $\mathbb{C}$), define

$$(I + a) + (I + b) = I + (a + b),$$
$$(I + a) \cdot (I + b) = I + ab,$$

and

$$\alpha(I + a) = I + \alpha a.$$

Then the operations above are well defined and under these operations A/I

is an algebra with unit $I + 1$. Further, define $\|I + a\| := \inf\{\|x + a\|: x \in I\}$ for a in A. Then A/I is a normed algebra with respect to this norm. If A is Banach, then A/I is Banach.

Proof. Straightforward.

Lemma 1.2.5 Let I be a proper ideal in a (real or complex) commutative Banach algebra A with unit 1. Then the closure J of I is a proper ideal. In particular, every maximal ideal is closed.

Proof. Clearly, J is an ideal. If $a \in I$ and $\|1 - a\| < 1$, then a is invertible by Lemma 1.1.14. Thus $\|1 - a\| \geq 1$ for every a in I and hence for every a in J. Hence $1 \notin J$, so that J is proper. $\qquad\square$

Theorem 1.2.6 Let A be a real commutative Banach algebra with unit 1.
(i) If $\phi \in \mathrm{Car}(A)$, then the *kernel of* ϕ, denoted by $\ker \phi := \{a \in A : \phi(a) = 0\}$, is a maximal ideal in A.
(ii) If M is a maximal ideal in A, then $M = \ker \phi$ for some ϕ in $\mathrm{Car}(A)$.
(iii) For ϕ, ψ in $\mathrm{Car}(A)$, if $\ker \phi = \ker \psi$, then $\psi = \phi$ or $\psi = \bar{\phi}$.

Proof.
   (i) If $a \in \ker \phi$, $b \in A$, then $\phi(ab) = \phi(a)\phi(b) = 0$. Thus $\ker \phi$ is an ideal. Since $\phi(1) = 1$, $\ker \phi$ is proper. Suppose that $\ker \phi$ is properly contained in an ideal I. We consider two cases to show that $I = A$.
   Case (a): Suppose that $\phi(A) \subset \mathbb{R}$. Then we can find b in I such that $\phi(b) = s \neq 0$, $s \in \mathbb{R}$. Let $a \in A$. If $c = a - \phi(a)b/s$, then $\phi(c) = 0$. Hence $c \in \ker \phi \subset I$. Thus $a = c + \phi(a)b/s \in I$. This shows that $I = A$.
   Case (b): If $\phi(A)$ is not contained in $\mathbb{R}$, we can find b in A such that $\phi(b) = s + it$ and $t \neq 0$. Let $c = (b - s)/t$. Then $\phi(c) = i$. There exists d in I such that $\phi(d) = p + iq \neq 0$, p, q in $\mathbb{R}$. Let $e = d(-pc^2 - qc)/(p^2 + q^2)$. Then $e \in I$ and $\phi(e) = 1$. Now let $a \in A$ and $\phi(a) = x + iy$. Then $a - xe - yce \in \ker \phi \subset I$. Also, $xe + yce \in I$. Hence $a \in I$. This shows that $I = A$.
   (ii) Let M be a maximal ideal. Then M is closed by Lemma 1.2.5. Lemma 1.2.4 shows that A/M is a real commutative Banach algebra with unit. Since M is maximal, it is easy to prove that A/M is a division algebra. Hence by the Mazur–Gelfand theorem 1.1.23, A/M is isomorphic to $\mathbb{R}$ or $\mathbb{C}$. Let θ be such an isomorphism and let $\pi: A \to A/M$ be the canonical map. Then $\phi := \theta \circ \pi \in \mathrm{Car}(A)$ and $\ker \phi = M$.
   (iii) Let $\phi, \psi \in \mathrm{Car}(A)$ and $\ker \phi = \ker \psi = M$. Let $\pi: A \to A/M$ be the canonical map and $\tilde{\phi}$ and $\tilde{\psi}$ be defined by $\tilde{\phi}((M + a)) := \phi(a)$ and $\tilde{\psi}((M + a)) := \psi(a)$ for a in A. Then $\tilde{\phi}$ and $\tilde{\psi}$ are both isomorphisms

from A/M to $\mathbb{R}$ or $\mathbb{C}$. Hence $\tilde{\phi} \circ (\tilde{\psi})^{-1}$ is an isomorphism from $\mathbb{R}$ to $\mathbb{R}$ or from $\mathbb{C}$ to $\mathbb{C}$. The only isomorphism from $\mathbb{R}$ to $\mathbb{R}$ is the identity map and the only isomorphisms from $\mathbb{C}$ to $\mathbb{C}$, where $\mathbb{C}$ is regarded as a real algebra, are (i) the identity map and (ii) the complex conjugation, that is, the map $z \mapsto \bar{z}$. This implies that $\tilde{\psi} = \tilde{\phi}$ or $\tilde{\psi} = \bar{\tilde{\phi}}$. Since $\phi = \tilde{\phi} \circ \pi$ and $\psi = \tilde{\psi} \circ \pi$, we have $\psi = \phi$ or $\psi = \bar{\phi}$.   □

Remark 1.2.7 The set of all maximal ideals of a real commutative Banach algebra A with unit is denoted by $M(A)$ and is called *the maximal ideal space of A*. Theorem 1.2.6 shows that the map ker: $\mathrm{Car}(A) \to M(A)$ (i.e., the map $\phi \mapsto \ker \phi$) is onto. This implies that $\mathrm{Car}(A)$ is nonempty. Further, for M in $M(A)$, the inverse image $\ker^{-1}(M)$ consists of $\{\phi, \bar{\phi}\}$ for some ϕ in $\mathrm{Car}(A)$.

Remark 1.2.8 Let A be a complex commutative Banach algebra with unit 1. The *carrier space of A*, denoted by $\mathrm{Car}(A)$, is the set of all nonzero homomorphisms from A to $\mathbb{C}$, where $\mathbb{C}$ is regarded as a complex algebra. Hence if $\phi \in \mathrm{Car}(A)$, then $\phi(i) = i$. Let A_R denote A, regarded as a real algebra. It is easy to see that if $\phi \in \mathrm{Car}(A)$, then ϕ as well as $\bar{\phi}$ are in $\mathrm{Car}(A_R)$. On the other hand, if $\phi \in \mathrm{Car}(A_R)$, then since $\phi(i)^2 = -1$, $\phi(i) = \pm i$. Thus exactly one of ϕ and $\bar{\phi}$ belongs to $\mathrm{Car}(A)$: that is, $\mathrm{Car}(A_R) = \mathrm{Car}(A) \cup \{\bar{\phi} : \phi \in \mathrm{Car}(A)\}$. In other words, $\mathrm{Car}(A_R)$ is a disjoint union of two copies of $\mathrm{Car}(A)$. Hence if $\phi \in \mathrm{Car}(A)$, then $\|\phi\| = 1$ by Lemma 1.2.2. Further, it is easy to see that the ideals of A and A_R are the same. Hence $M(A) = M(A_R)$. Thus statements (i) and (ii) of Theorem 1.2.6 remain valid for a complex algebra. As for statement (iii), by Remark 1.2.7, for M in $M(A)$, $\ker^{-1}(M)$ contains the points ϕ and $\bar{\phi}$ in $\mathrm{Car}(A_R)$. But exactly one of these is in $\mathrm{Car}(A)$. Thus for complex algebras, statement (iii) should be replaced by:

    (iii)′ If for ϕ, ψ in $\mathrm{Car}(A)$, $\ker \phi = \ker \psi$, then $\psi = \phi$.

    Hence the map ker: $\mathrm{Car}(A) \to M(A)$ is a bijection. This map is often used to identify the carrier space and the maximal ideal space of a complex commutative Banach algebra with unit element.

    We have now developed sufficient machinery to be able to present the main theorem.

Theorem 1.2.9 (Ingelstam) Let A be a real commutative Banach algebra with unit 1. Then
(i) $\mathrm{Car}(A)$, endowed with the Gelfand topology, is a compact Hausdorff space.

(ii) The map $\tau: \text{Car}(A) \rightarrow \text{Car}(A)$, defined by $\tau(\phi) = \bar{\phi}$, is a homeomorphism and $\tau \circ \tau$ is the identity map on $\text{Car}(A)$.
(iii) The set of all fixed points of τ is closed in $\text{Car}(A)$.
(iv) The mapping $a \mapsto \hat{a}$, $a \in A$, is a homomorphism of A into $C(\text{Car}(A))$, the algebra of all complex-valued continuous functions on $\text{Car}(A)$.
(v) a is singular in A if and only if $\hat{a}(\phi) = 0$ for some ϕ in $\text{Car}(A)$.
(vi) For $a \in A$, $\hat{a}(\text{Car}(A)) = \text{Sp}(a,A)$.
(vii) $\|\hat{a}\| := \sup\{|\hat{a}(\phi)| : \phi \in \text{Car}(A)\} = r(a)$.

Proof.
(i) Let $\phi, \psi \in \text{Car}(A)$ and $\phi \neq \psi$. This means that for some a in A, $\phi(a) \neq \psi(a)$; that is, $\hat{a}(\phi) \neq \hat{a}(\psi)$. Since $\hat{a}$ is continuous with respect to the Gelfand topology, we can enclose ϕ and ψ in two disjoint open sets, namely,

$$N_\phi := \{\theta \in \text{Car}(A) : |\hat{a}(\phi) - \hat{a}(\theta)| < \varepsilon\}$$

and

$$N_\psi := \{\theta \in \text{Car}(A) : |\hat{a}(\psi) - \hat{a}(\theta)| < \varepsilon\},$$

with $0 < \varepsilon < |\phi(a) - \psi(a)|2$. Thus $\text{Car}(A)$ is Hausdorff. To prove the compactness, we proceed as follows. For each a in A, let $K_A := \{z \in \mathbb{C} : |z| \leq \|a\|\}$ and $K := \Pi_{a \in A} K_a$ be the topological product. Since each K_a is compact (with the usual topology in $\mathbb{C}$), K is compact by Tychonoff's theorem [see Simmons (1963)]. We have observed in Lemma 1.2.2 that for ϕ in $\text{Car}(A)$ and a in A, $|\phi(a)| \leq \|a\|$. Thus $\phi(a) \in K_a$. Hence ϕ can be considered as a point of K. Consequently, $\text{Car}(A)$ can be considered as a subset of K. Further, it is clear that the Gelfand topology on $\text{Car}(A)$ is the same as the (relative) topology of $\text{Car}(A)$ as a subspace of K. Thus to prove the compactness of $\text{Car}(A)$, it is sufficient to show that $\text{Car}(A)$ is closed in K. Let $\psi \in K$ be in the closure of $\text{Car}(A)$. Since $\psi \in K$, ψ is a function on A such that $|\psi(a)| \leq \|a\|$ for each a in A. Since ψ is in the closure of $\text{Car}(A)$, there is a net $\{\psi_\alpha\}$ in $\text{Car}(A)$ converging to ψ. This means that $\psi_\alpha(a) \rightarrow \psi(a)$ for every a in A. Hence, for a,b in A,

$$\psi(a + b) = \lim \psi_\alpha(a + b) = \lim(\psi_\alpha(a) + \psi_\alpha(b)) = \psi(a) + \psi(b).$$

Similarly, $\psi(sa) = s\psi(a)$, $\psi(ab) = \psi(a)\psi(b)$, and $\psi(1) = 1$ for a,b in A, s in $\mathbb{R}$. Thus $\psi \in \text{Car}(A)$.

(ii) Clearly, $(\tau \circ \tau)(\phi) = \tau(\bar{\phi}) = \phi$ for all ϕ in $\text{Car}(A)$. Now suppose that a net $\{\phi_\alpha\}$ converges to ϕ in $\text{Car}(A)$. Then for each a in A, $\hat{a}(\phi_\alpha) \rightarrow \hat{a}(\phi)$. Hence $\phi_\alpha(a) \rightarrow \phi(a)$, or $\overline{\phi_\alpha(a)} \rightarrow \overline{\phi(a)}$; that is, $\hat{a}(\tau(\phi_\alpha)) \rightarrow \hat{a}(\tau(\phi))$. This means that $\tau(\phi_\alpha) \rightarrow \tau(\phi)$. Thus τ is continuous. Since $\tau^{-1} = \tau$, τ is a homeomorphism.

(iii) Follows from (ii).

(iv) Let $a, b \in A$ and $s \in \mathbb{R}$. For each ϕ in Car(A), we have $\phi(a + b) = \phi(a) + \phi(b)$; that is, $(a + b)^\wedge = \hat{a} + \hat{b}$. Similarly, $(sa)^\wedge = s\hat{a}$ and $(ab)^\wedge = \hat{a}\hat{b}$.

(v) If a is invertible, then $\phi(a)\phi(a^{-1}) = 1$. Hence $\phi(a) \neq 0$ for every ϕ in Car(A). If a is a singular, then $I := \{ab : b \in A\}$ is a proper ideal. I is contained in a maximal ideal M. By Theorem 1.2.6, $M = \ker \phi$ for some ϕ in Car(A). For this ϕ, $\phi(a) = 0 = \hat{a}(\phi)$.

(vi) $s + it \in \hat{a}(\text{Car}(A))$ if and only if $s + it = \phi(a)$ for some $\phi \in$ Car(A) if and only if $\phi((a - s)^2 + t^2) = 0$ for some ϕ in Car(A) if and only if $(a - s)^2 + t^2$ is singular in A if and only if $s + it \in \text{Sp}(a, A)$.

(vii) Follows from (vi) and the spectral radius formula (Theorem 1.1.19). $\qquad\square$

Remark 1.2.10 Let A be a complex commutative Banach algebra with unit 1. For $a \in A$, the *Gelfand transform* $\hat{a}$ *of* a is defined by $\hat{a}(\phi) = \phi(a)$ for ϕ in Car(A). The weakest topology on Car(A) that makes $\hat{a}$ continuous for each a in A is called the *Gelfand topology* on Car(A). Then:

(i) Car(A) is a compact Hausdorff space with respect to the Gelfand topology.

(ii) The mapping $a \mapsto \hat{a}$ is a homomorphism of A into $C(\text{Car}(A))$.

(iii) a is singular in A if and only if $\hat{a}(\phi) = 0$ for some ϕ in Car(A).

(iv) $\hat{a}(\text{Car}(A)) = \text{Sp}(a, A)$ for each a in A.

(v) $\|\hat{a}\| := \sup\{|\hat{a}(\phi)| : \phi \in \text{Car}(A)\} = r(a)$.

All the statements above can be proved easily from Theorem 1.2.9 by noting the following facts. If A_R denotes A, regarded as a real algebra, then by Remark 1.2.8,

$$\text{Car}(A_R) = \text{Car}(A) \cup \{\overline{\phi} : \phi \in \text{Car}(A)\}.$$

For a in A, the Gelfand transform $\hat{a}$ is the restriction to Car(A) of the Gelfand transform of a as an element of A_R. The latter is defined on Car(A_R). Also, by Remark 1.1.11,

$$\text{Sp}(a, A_R) = \text{Sp}(a, A) \cup \{\overline{\lambda} : \lambda \in \text{Sp}(a, A)\}.$$

Remark 1.2.11 Let A be a real commutative Banach algebra with unit 1. Since the map ker: Car(A) $\to M(A)$ is onto, we can consider the quotient topology on $M(A)$: A subset $G \subset M(A)$ is defined to be open if $\ker^{-1}(G)$ is open in Car(A). Then it is easy to show that $M(A)$ is a compact Hausdorff space with respect to this topology. We shall call this topology the *Gelfand topology in* $M(A)$. The map ker is obviously continuous. It is also open because if G is an open subset of Car(A), then $G \cup$

$\tau(G)$ is open. But $G \cup \tau(G) = \ker^{-1}(\ker G)$, so that $\ker G$ is open in $M(A)$. Essentially, $M(A)$ is the quotient space obtained from $\text{Car}(A)$ by identifying ϕ with $\tau(\phi)$. Now let $a \in A$. Since $\hat{a}$ is a continuous complex-valued function on $\text{Car}(A)$, $\text{Re } \hat{a}$ and $|\hat{a}|$ are real-valued continuous functions on $\text{Car}(A)$. By Remark 1.2.7, the functions $\text{Re } \hat{a}$ and $|\hat{a}|$ take constant values on $\ker^{-1}(M)$ for any M in $M(A)$. Hence they induce the continuous maps $\text{Re } \hat{a} \colon M(A) \to \mathbb{R}$ and $|\hat{a}| \colon M(A) \to \mathbb{R}$ defined by $\text{Re } \hat{a}(M) = \text{Re } \phi(a)$ and $|\hat{a}|(M) = |\phi(a)|$, where $M \in M(A)$ and $\ker \phi = M$.

Let $\text{Re } \hat{A} := \{\text{Re } \hat{a} \colon a \in A\}$ and $|\hat{A}| := \{|\hat{a}| \colon a \in A\}$. It was shown in Limaye (1976) that the weak $\text{Re } \hat{A}$ topology on $M(A)$ [i.e., the smallest topology on $M(A)$ making $\text{Re } \hat{a}$ continuous for all a in A] is the same as the weak $|\hat{A}|$ topology on $M(A)$. In fact, this topology is the same as the quotient topology defined above. To see this, consider a net $\{M_\alpha\}$ and an element M in $M(A)$. Let ϕ_α, ϕ be in $\text{Car}(A)$ such that $\ker \phi_\alpha = M_\alpha$ for each α and $\ker \phi = M$. Now if G is any neighborhood of M, then $\ker^{-1}(G)$ contains ϕ as well as $\overline{\phi}$. Thus it is easy to check that $\{M_\alpha\}$ converges to M in the quotient topology if and only if

(i) $\min\{|\hat{a}(\phi_\alpha) - \hat{a}(\phi)|, |\hat{a}(\phi_\alpha) - \hat{a}(\overline{\phi})|\} \to 0$ for every a in A.

Similarly, M_α converges to M in the weak $\text{Re } \hat{A}$ (respectively, $|\hat{A}|$) topology if and only if:

(ii) $\text{Re } \hat{a}(\phi_\alpha) \to \text{Re } \hat{a}(\phi)$ for every a in A [respectively,

(iii) $|\hat{a}|(\phi_\alpha) \to |\hat{a}|(\phi)$ for every a in A].

We now show that statements (i), (ii), and (iii) are equivalent for every net $\{\phi_\alpha\}$ in $\text{Car}(A)$. This implies that all the three topologies on $M(A)$ are the same. (i) obviously implies (ii) and (iii). "(ii) implies (iii)" follows from the relation

$$|\hat{a}|^2 = 2(\text{Re } \hat{a})^2 - \text{Re}(a^2)\hat{} \qquad \text{for every } a \text{ in } A.$$

Similarly, "(iii) implies (ii)" follows by noting that

$$\text{Re } \hat{a} = \log|(\exp(a))\hat{}| \qquad \text{for every } a \text{ in } A.$$

Thus (ii) and (iii) are equivalent. We shall now show that (ii) and (iii) imply (i). Let $a \in A$, $\hat{a}(\phi_\alpha) = x_\alpha + iy_\alpha$, and $\hat{a}(\phi) = x + iy$. (ii) and (iii) imply that $x_\alpha \to x$ and $x_\alpha^2 + y_\alpha^2 \to x^2 + y^2$. Hence $y_\alpha^2 \to y^2$; that is, $|y_\alpha| \to |y|$. Now

$$\min\{|\hat{a}(\phi_\alpha) - \hat{a}(\phi)|^2, \ |\hat{a}(\phi_\alpha) - \hat{a}(\overline{\phi})|^2\}$$
$$= \min\{(x_\alpha - x)^2 + (y_\alpha - y)^2, \ (x_\alpha - x)^2 + (y_\alpha + y)^2\}$$
$$= (x_\alpha - x)^2 + \min\{(y_\alpha - y)^2, \ (y_\alpha + y)^2\}$$
$$= (x_\alpha - x)^2 + (|y_\alpha| - |y|)^2 \to 0.$$

Corollary 1.2.12 Let A be a real or complex commutative Banach algebra with unit 1. Then the following statements are equivalent.
(i) $\|a\|^2 = \|a^2\|$ for all a in A.
(ii) $r(a) = \|a\|$ for all a in A.
(iii) The Gelfand mapping $a \mapsto \hat{a}$ is an isometric isomorphism from A onto a closed subalgebra of $C(\text{Car}(A))$.

Proof.
(i) implies (ii): By induction, $\|a^{2^m}\| = \|a\|^{2^m}$ for $m = 1,2,\dots$. Hence, by Lemma 1.1.13,

$$r(a) = \lim_{n \to \infty} \|a^n\|^{1/n}$$

$$= \lim_{m \to \infty} \|a^{2^m}\|^{1/2^m}$$

$$= \|a\|.$$

(ii) implies (iii): By Theorem 1.2.9 and Remark 1.2.10, $\hat{A} := \{\hat{a} : a \in A\}$ is a subalgebra, the map $a \mapsto \hat{a}$ is a homomorphism, and $\|\hat{a}\| = r(a) = \|a\|$. Since A is Banach, $\hat{A}$ is closed in $C(\text{Car}(A))$.

(iii) implies (i):

$$\|a^2\| = \|(a^2)\hat{}\| = \|(\hat{a})^2\|$$

$$= \|\hat{a}\|^2 \qquad \text{by definition of } \|\hat{a}\|$$

$$= \|a\|^2. \qquad\qquad \square$$

Every complex Banach algebra is also a real Banach algebra. To distinguish between complex Banach algebras and those real Banach algebras that behave in a different manner, Ingelstam (1964) introduced and studied some "reality conditions." These conditions indicate the extent to which a real Banach algebra is different from complex algebras. The strongest among these conditions was called "strict reality."

A real Banach algebra A with unit 1 is called *strictly real* if $1 + a^2$ is invertible for all a in A. It is easy to see that if A is strictly real, then $\text{Sp}(a) \subset \mathbb{R}$ for all a in A. In fact, if $a \in A$ and $s + it \in \text{Sp}(a)$ with $t \neq 0$, then $(a - s)^2 + t^2$ is singular, that is, $1 + ((a - s)/t)^2$ is singular, in contradiction to the strict reality of A.

Now suppose that A is commutative and strictly real. Then Theorem 1.2.9(vi) implies that the Gelfand transform $\hat{a}$ is real for each a in A. Thus $\phi = \bar{\phi}$ for all ϕ in $\text{Car}(A)$. As $\hat{A}$ is a real subalgebra of $C_{\mathbb{R}}(\text{Car}(A))$ that contains the constant function $\hat{1}$ and separates the points of $\text{Car}(A)$, $\hat{A}$ is dense in $C_{\mathbb{R}}(\text{Car}(A))$ by the real Stone–Weierstrass theorem. (See also Corollary 2.1.4 and the comments following it.)

If A is real Banach algebra with 1 satisfying $\|a\|^2 \le k\|a^2 + b^2\|$ for all a,b in A with $ab = ba$ and some positive constant k, then A is commutative by Theorem 1.1.30. It can also be proved that $\text{Sp}(a) \subset \mathbb{R}$ for each a in A [see Kulkarni and Limaye (1980, Theorem 2.4)]. Hence $\hat{A}$ is dense in $C_\mathbb{R}(\text{Car}(A))$. Next, by Lemma 1.1.29(ii) and Theorem 1.2.9(vii), $\|a\| \le kr(a) = k\|\hat{a}\|$ for each a in A. Hence $\hat{A}$ is uniformly closed; that is, $\hat{A} = C_\mathbb{R}(\text{Car}(A))$. Thus A is isomorphic and homeomorphic to $C_\mathbb{R}(X)$ for some compact Hausdorff space. This is a well-known theorem of Arens (Arens and Kaplansky, 1948).

1.3 REAL FUNCTION ALGEBRAS

We begin this section with the study of a few properties of the complex Banach algebra $C(X)$, described in Example 1.1.2. Recall that this is the algebra of all complex-valued continuous functions on a compact Hausdorff space, with pointwise operations and the supremum norm. Frequently, we shall need to take the supremum over subsets of X.

Notation For a nonempty subset K of X and f in $C(X)$, we denote

$$\|f\|_K := \sup\{|f(x)| : x \in K\}.$$

The subscript K will be omitted when $K = X$.

Definition 1.3.1 Let X be a compact Hausdorff space and A be a nonempty subset of $C(X)$. For each x in X, *the evaluation map at x*, denoted by e_x, is defined by

$$e_x(f) = f(x) \qquad \text{for } f \in A.$$

It is easily seen that if A is a subspace, then $e_x : A \to \mathbb{C}$ is a linear map, and if A is a subalgebra, then e_x is a homomorphism. If A contains 1, then $e_x(1) = 1$ and hence $e_x \ne 0$. The next theorem shows that all the nonzero homomorphisms of $C(X)$ to $\mathbb{C}$ are of this form.

Theorem 1.3.2 Let X be a compact Hausdorff space. Then $\text{Car}(C(X)) = \{e_x : x \in X\}$.

Proof. We have already observed that $e_x \in \text{Car}(C(X))$ for each x in X. Now let $\phi \in \text{Car}((C(X))$. By Remark 1.2.8, it is enough to prove that

$$\ker \phi = \ker e_x \qquad \text{for some } x \text{ in } X.$$

For f in $C(X)$, let $Z(f) := \{y \in X : f(y) = 0\}$. We claim that

$$K := \cap \{Z(f) : f \in \ker \phi\} \text{ is nonempty.}$$

Since each $Z(f)$ is a closed subset, by the finite intersection property of the compact space X, it is sufficient to prove that for $f_1,\ldots,f_n$ in ker ϕ, $Z(f_1) \cap \cdots \cap Z(f_n)$ is nonempty.

If possible, let $Z(f_1) \cap \cdots \cap Z(f_n)$ be empty. Then $g := \Sigma_{i=1}^n f_i \bar{f}_i \in$ ker ϕ. But since f_i's have no common zero, $g(y) > 0$ for all y in X. Hence g is invertible, so that $\phi(g) \neq 0$. This contradiction proves the claim.

Now $I := \{f \in C(X) : f \equiv 0 \text{ on } K\}$ is a proper ideal of $C(X)$. By the claim above, ker $\phi \subset I$. Since ker ϕ is a maximal ideal, we have ker $\phi = I$ and K is a singleton, say $K = \{x\}$. $\qquad\square$

Corollary 1.3.3 Let $(C(X))_R$ denote $C(X)$, regarded as a real algebra. Then $\mathrm{Car}((C(X))_R) = \{e_x : x \in X\} \cup \{\bar{e}_x : x \in X\}$.

Proof. Follows from Theorem 1.3.2 and Remark 1.2.8.

Definition 1.3.4 (Involutions) Let X be a topological space. A map $\tau \colon X \to X$ is called a *topological involution* on X if τ is a homeomorphism and $\tau(\tau(x)) = x$ for all x in X. (Note that, a priori, we could merely assume that τ is continuous.)

Let A be a real or complex vector space. A *linear involution* on A is a map $a \mapsto a^*$ of A into A satisfying the following axioms: For a,b in A and a scalar α:

(i) $(a + b)^* = a^* + b^*$.
(ii) $(\alpha a)^* = \bar{\alpha} a^*$.
(iii) $(a^*)^* = a$.

The scalar α can be a real or complex number depending on whether A is a real or complex vector space. Note that if A is a complex vector space, then a linear involution on A is *not* a linear map; but it is conjugate linear.

Now let A be an algebra over $\mathbb{R}$ or $\mathbb{C}$. An *algebra involution* on A is a linear involution on A satisfying $(ab)^* = b^* a^*$ for all a,b in A.

The adjective "topological," "linear," or "algebra" will be omitted when the type of an involution is clear from the context.

Let X be a compact Hausdorff space and τ a topological involution on X. We now take a closer look at the algebra $C(X,\tau)$ defined in Example 1.1.2(b). Recall that

$$C(X,\tau) := \{f \in C(X) : f(\tau(x)) = \bar{f}(x) \text{ for all } x \in X\}.$$

This algebra was defined explicitly in Kulkarni and Limaye (1981b). However, a careful reading of the papers of Arens and Kaplansky (1948) and Ingelstam (1964) will reveal the mention of this algebra in their papers. The main object of the present monograph is to investigate $C(X,\tau)$ and some of its subalgebras. One may ask: "Why should we study $C(X,\tau)$?"

The algebras $C(X)$ and $C_R(X)$ are well known in the literature. Many of their properties have been thoroughly investigated. We have already observed that if τ is the identity map on X, then $C(X,\tau)$ reduces to $C_R(X)$. On the other hand, $C(X)$ can also be treated as $C(Y,\tau)$, where Y is a disjoint union of two copies of X and τ is a homeomorphism that sends a point in one copy of X to the corresponding point in the other copy. [Precisely, $Y = X \times \{0,1\}$ and $\tau((x,j)) = (x, (j + 1)$ modulo 2) for $(x,j) \in Y$.] Identifying X with one of the copies (say, $X \times \{0\}$), X can be treated as a subspace of Y. Then every f in $C(X)$ can be extended uniquely to Y by requiring that the extension belongs to $C(Y,\tau)$.

Thus $C(X,\tau)$ is a more general object than $C_R(X)$ and $C(X)$. Hence every result that we prove about $C(X,\tau)$ will be valid for $C_R(X)$ and $C(X)$. On the other hand, it should be of interest to know how many properties of these better known algebras are shared by the algebra $C(X,\tau)$.

Theorem 1.3.5 Let X be a compact Hausdorff space and τ a topological involution on X. Define $\sigma: C(X) \to C(X)$ by $\sigma(f)(x) = \bar{f}(\tau(x))$ for $f \in C(X)$, $x \in X$. [In short, $\sigma(f) = \bar{f} \circ \tau$.] Then:
(i) σ is an algebra involution on $C(X)$ and

$$C(X,\tau) = \{f \in C(X): \sigma(f) = f\}.$$

(ii) $C(X) = C(X,\tau) \oplus iC(X,\tau)$; that is, every h in $C(X)$ can be expressed uniquely as $f + ig$ with f, g in $C(X,\tau)$.
(iii) σ is an isometry.
(iv) For f in $C(X)$, define $P(f) = [f + \sigma(f)]/2$. Then P is a continuous linear projection; that is, P is a continuous linear map satisfying $P^2 = P$ of $(C(X))_R$ onto $C(X,\tau)$.
(v) Every algebra involution on $C(X)$ arises from a topological involution on X in the manner described above.

Proof.
    (i) Obvious.
    (ii) Since σ is an algebra involution, for f in $C(X)$, $\sigma(f + \sigma(f)) = f + \sigma(f)$ and $\sigma((f - \sigma(f))/i) = (f - \sigma(f))/i$. Thus $f = (f + \sigma(f))/2 + i((f - \sigma(f))/2i)$ with $(f + \sigma(f))/2$, $(f - \sigma(f))/2i$ in $C(X,\tau)$. Further, if $h = f + ig$ with f and g in $C(X,\tau)$, then $\sigma(h) = f - ig$ and hence $f = (h + \sigma(h))/2$ and $g = (h - \sigma(h))/2i$. This proves the uniqueness of f and g.
    (iii) Let $f \in C(X)$. Then

$$\|\sigma(f)\| = \sup\{|\bar{f}(\tau(x))| : x \in X\}$$
$$= \sup\{|f(y)| : y \in X\}, \qquad \text{because } \tau \text{ is an involution on } X$$
$$= \|f\|.$$

(iv) follows from (i) and (iii).

(v) Let σ be an algebra involution on $C(X)$. We prove that σ is induced by a topological involution τ on X. Let $x \in X$. Define $\phi_x: C(X) \to \mathbb{C}$ by $\phi_x(f) = \overline{\sigma(f)}(x)$, $f \in C(X)$. It is routine to check that $\phi_x \in \text{Car}(C(X))$. Hence, by Theorem 1.3.2, $\phi_x = e_y$ for a unique y in X. We define $y = \tau(x)$. Thus we have $f(\tau(x)) = \overline{\sigma(f)}(x)$ for all f in $C(X)$, x in X. Replacing x by $\tau(x)$, $f(\tau(\tau(x))) = \overline{\sigma(f)}(\tau(x)) = \sigma((\overline{\sigma(f)}))(x) = f(x)$ for all f in $C(X)$. Hence $\tau(\tau(x)) = x$ for all x in X. Thus it remains only to show that τ is continuous. First note that since $\overline{\sigma(f)}(x) = f(\tau(x))$ for all f in $C(X)$, x in X, and $\tau \circ \tau$ is the identity map, we have $\|\sigma(f)\| = \|f\|$. Hence σ is continuous. Now suppose that a net $\{x_\alpha\}$ converges to a point x in X. Then for every f in $C(X)$, $f(x_\alpha) \to f(x)$. Since σ is continuous, this implies that $\sigma(f)(x_\alpha) \to \sigma(f)(x)$. Hence $f(\tau(x_\alpha)) = \overline{\sigma(f)}(x_\alpha) \to \overline{\sigma(f)}(x) = f(\tau(x))$. This means that $\tau(x_\alpha) \to \tau(x)$; that is, τ is continuous on X. $\qquad\square$

Note that $\tau = $ the identity map on X gives rise to $\sigma(f) = \bar{f}$ and in this case $C(X) = C_\mathbb{R}(X) + iC_\mathbb{R}(X)$.

In proving assertion (v) of Theorem 1.3.5, we have used the following properties of $C(X)$:

(i)   If for x,y in X, $f(x) = f(y)$ for all f in $C(X)$, then $x = y$.
(ii)   If for a net $\{x_\alpha\}$ and a point x in X, $f(x_\alpha) \to f(x)$ for all f in $C(X)$, then $x_\alpha \to x$.

These facts follow from Urysohn's lemma. Because X is a compact Hausdorff space, if $x,y \in X$ and $x \neq y$, we can find a continuous function $f: X \to [0,1]$ such that $f(x) = 1$ and $f(y) = 0$.

Definition 1.3.6   Let A be a set of functions on a set X. We say that A *separates the points of* X if for x,y in X and $x \neq y$, there exists f in A such that $f(x) \neq f(y)$.

Let τ be an involution on X and E be a set of functions on X such that $u \circ \tau = u$ for every u in E. We say that E *separates the points of* X/τ if for x,y in X and $x \neq y$, $\tau(x) \neq y$, there exists u in E such that $u(x) \neq u(y)$.

Lemma 1.3.7   Let X be a compact Hausdorff space and τ a topological involution on X. Let $x,y \in X$ and $x \neq y$.
(i)   If $y = \tau(x)$, then there exists f in $C(X,\tau)$ such that $f(x) = i$ and $f(y) = -i$.

(ii) If $y \neq \tau(x)$, then there exists f in $C(X,\tau)$ such that $f(x) = 1$ and $f(y) = 0$.

In particular, $C(X,\tau)$ separates the points of X. and $\mathrm{Re}\ C(X,\tau)$ separates the points of X/τ.

Proof. As in Theorem 1.3.5, for h in $C(X)$, define $\sigma(h) = \bar{h} \circ \tau$. Then, by the same theorem, $\sigma(h + \sigma(h)) = h + \sigma(h)$ and $\sigma(h\sigma(h)) = h\sigma(h)$. Hence $h + \sigma(h)$ and $h\sigma(h) \in C(X,\tau)$.

(i) We can find h in $C(X)$ such that $h(x) = i$ and $h(y) = 0$. Let $f = h + \sigma(h)$. Then $f \in C(X,\tau)$, $f(x) = i$, and $f(y) = -i$.

(ii) We can find h in $C(X)$ such that $h(x) = h(\tau(x)) = 1$ and $h(y) = 0$. Let $f = h\sigma(h)$. Then $f \in C(X,\tau)$, $f(x) = 1$, and $f(y) = 0$. $\square$

We now give the definition of a mathematical structure that is our main concern in this book.

Definition 1.3.8 Let X be a compact Hausdorff space and τ a topological involution on X. A *real function algebra on* (X,τ) is a real subalgebra A of $C(X,\tau)$ such that:
(i) A is uniformly closed.
(ii) $1 \in A$.
(iii) A separates the points of X.

A *complex function algebra* on a compact Hausdorff space X is a complex subalgebra of $C(X)$, satisfying conditions (i), (ii), and (iii).

Notice that if B is a complex subalgebra of $C(X)$ containing 1 and separating the points of X, we can assert that for x, y in X and $x \neq y$, there exists f in B such that $f(x) = 0$ and $f(y) = 1$. This can be seen as follows: There exists g in B such that $a := g(x) \neq g(y) := b$. Let $f := (g - a)/(b - a)$. Then $f \in B$, $f(x) = 0$, and $f(y) = 1$. However, we cannot make the assertion above if B is only a real subalgebra of $C(X)$. In particular, if τ is a topological involution on X, and B is a (real) subalgebra of $C(X,\tau)$, then for every $f \in B$, $f(x) = 0$ implies that $f(\tau(x)) = 0$.

Lemma 1.3.9 Let X and τ be as in Definition 1.3.8. Let A be a subalgebra of $C(X,\tau)$ that separates the points of X and contains 1. Let $x, y \in X$ with $x \neq y$. Then:
(i) If $y = \tau(x)$, there exists f in A such that $f(x) = i$ and $f(y) = -i$.
(ii) If $y \neq \tau(x)$, there exists f in A such that $f(x) = 1$ and $f(y) = 0$.
In particular, $\mathrm{Re}\ A$ separates the points of X/τ.

Proof. Since A separates the points of X, there exists f_1 in A such that $\alpha_1 := f_1(x) \neq f_1(y) =: \beta_1$.

(i) Let $y = \tau(x)$ and $\alpha_1 = a + ib$. Then $\beta_1 = a - ib$, so that $b \neq 0$. Now $f := (f_1 - a)/b \in A$, $f(x) = i$, and $f(y) = -i$.

(ii) Let $y \neq \tau(x)$. Then there is f_2 in A such that $\alpha_2 := f_2(\tau(x)) \neq f_2(y) := \beta_2$. Now consider

$$f := \left[1 - \frac{(f_1 - \alpha_1)(f_2 - \alpha_2)}{(\beta_1 - \alpha_1)(\beta_2 - \alpha_2)}\right]\left[1 - \frac{(f_1 - \bar{\alpha}_1)(f_2 - \bar{\alpha}_2)}{(\bar{\beta}_1 - \bar{\alpha}_1)(\bar{\beta}_2 - \bar{\alpha}_2)}\right].$$

Let $r := (\beta_1 - \alpha_1)(\beta_2 - \alpha_2)$. Then

$$f = 1 - \frac{2}{|r|^2}\left[\mathrm{Re}(\bar{\alpha}_1\bar{\alpha}_2 r) + \mathrm{Re}(\bar{\alpha}_2 r)f_1 + \mathrm{Re}(\bar{\alpha}_1 r)f_2 + \mathrm{Re}(r)f_1 f_2\right]$$

$$+ \frac{1}{|r|^2}[f_1^2 - 2\,\mathrm{Re}(\alpha_1)f_1 + |\alpha_1|^2][f_2^2 - 2\,\mathrm{Re}(\alpha_2)f_2 + |\alpha_2|^2].$$

Hence $f \in A$. Now it is easy to check that $f(x) = 1$ and $f(y) = 0$. $\square$

Like complex Banach algebras, complex function algebras have been objects of detailed investigation. Most books on Banach algebras contain some material devoted to complex function algebras. In addition, since the late 1960s several books dealing entirely with complex function algebras have appeared: for example, Browder (1969), Gamelin (1969), Stout (1971), and Suciu (1975). On the other hand, real function algebras are of recent origin. These algebras were defined for the first time in Kulkarni and Limaye (1981b). Since then, many mathematicians have attempted to develop the theory of real function algebras along the lines of complex function algebras. Our aim in this book is to present a systematic account of these attempts. We shall soon establish that the class of real function algebras includes that of complex function algebras. Hence every result about real function algebras can be regarded as a generalization of the corresponding result about complex function algebras.

Assume that condition (i) of Definition 1.3.8 is replaced by the following:

(i)′ A is a Banach algebra with respect to some norm.

Then the resulting algebra is known as a *real (or complex) Banach function algebra*. Obviously, every (real or complex) function algebra is a Banach function algebra. The Wiener algebra (Example 1.1.5) is a complex Banach function algebra, which is not a function algebra, because the norm defined there is not the supremum norm.

We now discuss a few examples of function algebras other than the trivial ones: For a compact Hausdorff space X, $C(X)$ is a complex function algebra, and if τ is a topological involution on X, then $C(X,\tau)$ is a real function algebra.

Example 1.3.11 Let X be a compact subset of $\mathbb{C}$ and B the algebra of all functions that are uniform limits of sequences of rational functions p/q, where p and q are polynomials and q has no zero on X. Then B is a $C(X)$ that are analytic in the open unit disk $\{z \in \mathbb{C}: |z| < 1\}$. Then B is a complex function algebra, known in the literature as the *disk algebra*. This algebra serves as a model for the development of the theory of complex function algebras. Now, define a map $\tau: X \to X$ by $\tau(z) = \bar{z}$ for $z \in X$. Let $A = \{f \in B: f(\tau(z)) = \overline{f(z)}\}$. Then A equals the set of all uniform limits of polynomials with real coefficients on X. A is a real function algebra. In fact, $A = B \cap C(X,\tau)$. A is called the *real disk algebra* and is useful as a model to develop the theory of real function algebras. We now prove that the carrier space of A can be identified with the disk X, that is,

$$\mathrm{Car}(A) = \{e_\lambda: \lambda \in X\}.$$

Clearly, if $\lambda \in X$, $e_\lambda \in \mathrm{Car}(A)$. To prove the reverse inclusion, consider $\phi \in \mathrm{Car}(A)$. Consider the polynomial z in A: z is the function defined as $z(\lambda) = \lambda$ for all λ in X. Let $\phi(z) = \lambda$. Since $\|z\| = 1$, we have $|\lambda| \leq 1$. Thus $\lambda \in X$. Since $\phi \in \mathrm{Car}(A)$, $\phi(z^n) = \lambda^n$ for $n = 1,2,\dots$. Hence $\phi(p) = p(\lambda)$ for every polynomial p with real coefficients and hence $\phi(f) = f(\lambda)$ for every $f \in A$. Thus $\phi = e_\lambda$.

Similarly, we can prove that $\mathrm{Car}(B) = \{e_\lambda: \lambda \in X\}$. (Note that we have used the same symbol, e_λ, for a homomorphism from B to $\mathbb{C}$.) Thus the maximal ideals of B are of the form

$$M_z = \{f \in B: f(z) = 0\},$$

where $z \in \mathbb{C}$ and $|z| \leq 1$.

The closed ideals of the disk algebra B can be described as follows (Rudin, 1957). Let J be a closed ideal in B. Then there is a bounded analytic function q on the open unit disk and a closed subset E of the unit circle such that the Lebesgue measure of E is zero,

$$\left|\lim_{r \to 1} q(re^{i\theta})\right| = 1 \qquad \text{for almost all } \theta,$$

$$J = \{qf: f \in B, f = 0 \text{ on } E\}.$$

More generally, we may consider the algebra B consisting of uniform limits of sequences of polynomials on a compact subset of $\mathbb{C}$, and the algebra A consisting of uniform limits of sequences of polynomials with real coefficients on a compact subset of $\mathbb{C}$ that is symmetric about the real axis (Definition 1.1.24).

Example   1.3.11   Let X be a compact subset of $\mathbb{C}$ and B the algebra of all functions that are uniform limits of sequences of rational functions p/q, where p and q are polynomials and q has no zero on X. Then B is a complex function algebra. Further, if X is symmetric about the real axis, we can define $\tau: X \to X$ by $\tau(z) = \bar{z}$, $z \in X$. Let $A := \{f \in B : f(\tau(z)) = \bar{f}(z)$ for all z in $X\}$. Then A is a real function algebra consisting of all uniform limits of sequences of rational functions p/q, where p and q are polynomials with real coefficients and q does not have any zero on X.

Example   1.3.12   Let $0 < c < d$ and X be the annular region $X := \{z \in \mathbb{C} : c \le |z| \le d\}$. Let

$$B := \{f \in C(X) : f \text{ is analytic in the interior of } X\}.$$

Then B is a complex function algebra on X. Now let $0 < c < 1$ and $d = 1/c$. Then we can define a map $\tau: X \to X$ by $\tau(z) = -1/\bar{z}$. This map is a topological involution on X. Define

$$A := \{f \in B : f(\tau(z)) = \bar{f}(z) \text{ for all } z \in X\}.$$

Then A is a real function algebra on (X, τ).

    More generally, an algebra B can similarly be defined on any compact Riemann surface with a nonempty boundary. [We refer the reader to Springer (1957) for an introduction to Riemann surfaces.] Now let Y be a compact nonorientable Klein surface with nonempty boundary ∂Y. Then Y admits an orienting double (X, p, τ), where X is a compact Riemann surface with boundary ∂X, $p: X \to Y$ is a 2:1 covering map such that $p(\partial X) = \partial Y$, $p^{-1}(\partial Y) = \partial X$, and $\tau: X \to X$ is an antianalytic involution of X such that $p \circ \tau = p$. [For definitions and proof, see Alling and Greenleaf (1971, p. 140).] Then we can define algebras B and A as above. A is called *the standard algebra on a Klein surface Y*. In the particular case mentioned above, the annular region X, with $d = 1/c$, is the orienting double of a Möbius strip. A description of the closed ideals of the function algebras B and A is given in Alling and Limaye (1972) on the lines of the description of the closed ideals of the disk algebra mentioned in Example 1.3.10.

    It follows from the comments at the end of Section 1.2 that if a real function algebra A on (X, τ) is strictly real, then τ is the identity map on X and $A = C_{\mathbb{R}}(X)$. Hwang (1990) has studied the implications of this and other reality conditions in the context of a real function algebra.

Definition 1.3.13 A *real uniform algebra* is a real commutative Banach algebra A with unit 1 such that $\|a\|^2 = \|a^2\|$ for all a in A.

Remark 1.3.14 One can similarly define a complex uniform algebra. In fact, in this case, the assumption of commutativity need not be made, a priori, as any complex Banach algebra B satisfying $\|a\|^2 = \|a^2\|$ for all $a \in B$ is commutative by Theorem 1.1.30(i). Example 1.1.4 shows that this theorem does not hold for real Banach algebras.

Real uniform algebras were first defined in Limaye and Simha (1975). Obviously, every real function algebra is a real uniform algebra. Here are some further examples of real uniform algebras.

Example 1.3.15 Let $D := \{z \in \mathbb{C} : |z| < 1\}$ be the open unit disk and A be the set of all complex-valued bounded analytic functions f on D satisfying $f(\bar{z}) = \overline{f(z)}$ for all $z \in D$. *With the supremum norm, A is obviously a real uniform algebra.*

Example 1.3.16 Let μ be a positive measure on a σ-algebra on a measure space X, and let $\tau : X \to X$ be a map such that $\tau(\tau(x)) = x$ for all x in X, and $\tau(E)$ is measurable whenever E is a measurable subset of X. [As a particular case, we may consider $X = \mathbb{R}$, μ the Lebesgue measure, and $\tau(x) = -x$ for x in $\mathbb{R}$.] Let

$$A := \{f \in L^\infty(\mu) : f(\tau(x)) = \bar{f}(x) \text{ for all } x \text{ in } X\}.$$

A is clearly a real uniform algebra.

We will now show that real uniform algebras are real function algebras in disguise. To penetrate this disguise, we have to take help of the Gelfand representation developed in Section 1.2.

Theorem 1.3.17 Every real uniform algebra is isometrically isomorphic to a real function algebra.

Proof. Let A be a real uniform algebra. Let $X = \text{Car}(A)$ and $\tau : X \to X$ be the map $\tau(\phi) = \bar{\phi}$. By Ingelstam's theorem (Theorem 1.2.9), X is a compact Hausdorff space and τ is a topological involution on X. Let $\hat{A} = \{\hat{a} : a \in A\}$, where $\hat{a}$ denotes the Gelfand transform of a. By Corollary 1.2.12, the Gelfand mapping $a \mapsto \hat{a}$ is an isometric isomorphism of A onto $\hat{A}$. Thus it suffices to prove that $\hat{A}$ is a real function algebra on (X,τ). Clearly, for each a in A and ϕ in X, $\hat{a}(\tau(\phi)) = \bar{\phi}(a) = \overline{\hat{a}(\phi)}$. Hence $\hat{a} \in C(X,\tau)$. Thus $\hat{A}$ is a real subalgebra of $C(X,\tau)$. It contains the constant function $\hat{1}$; it obviously separates the points of X. Since $\|\hat{a}\| = \|a\|$ for each a in A, $\hat{A}$ is uniformly closed. □

Note that we need not choose X to be the entire carrier space of A in proving Theorem 1.3.17. Instead, we could as well choose Y to be any closed subset of X satisfying $\tau(Y) = Y$ and

$$\sup\{|\phi(a)| : \phi \in X\} = \sup\{|\phi(a)| : \phi \in Y\}$$

for every a in A. Then $\hat{A}_{|Y} := \{\hat{a}|_Y : a \in A\}$ is a real function algebra on $(Y, \tau|_Y)$, which is isometrically isomorphic to A. A natural choice for such a set Y can often be made if we know something more about the algebra A. This happens for a class of real uniform algebras studied by Mehta and Vasavada (1986).

Corollary 1.3.18 Let X be a compact Hausdorff space and A be a uniformly closed real subalgebra of $C(X)$ containing constants and separating the points of X. For ϕ in $\text{Car}(A)$, let $\tau(\phi) = \bar{\phi}$. Identifying $x \in X$ with $e_x \in \text{Car}(A)$, let $Y = X \cup \tau(X) \subset \text{Car}(A)$. Then $\hat{A}_{|Y}$ is a real function algebra on $(Y, \tau|_Y)$, and A is isometrically isomorphic to $\hat{A}_{|Y}$.

Proof. Every ϕ in Y equals e_x or $\bar{e}_x$ for some x in X, so that for every f in A, $\hat{f}(\phi) = \hat{f}(\tau(\phi))$, and

$$\|\hat{f}_{|Y}\| = \sup\{|\hat{f}(\phi)| : \phi \in Y\} = \sup\{|f(x)| : x \in X\} = \|f\|.$$

Hence $\hat{A}_{|Y}$ is uniformly closed in $C(Y, \tau|_Y)$. Clearly, $\hat{1}_{|Y}$ is a constant function on Y and $\hat{A}_{|Y}$ separates the points of Y. Thus $\hat{A}_{|Y}$ is a real function algebra on $(Y, \tau|_Y)$. It is also obvious that the map $f \mapsto \hat{f}_{|Y}$ is an isometric isomorphism of A onto $\hat{A}_{|Y}$. □

Remark 1.3.19
(i) In Corollary 1.3.18, suppose that X and A have the following property: For each x in X, there is y in X such that $f(y) = \bar{f}(x)$ for all f in A. This means that $e_y = \bar{e}_x$; that is, $\tau(x) \in X$. Hence $Y = X$ and τ is an involution on X.

As an example of this case, we may consider

$$A := \{f \in C([0,1]) : f(1 - t) = \bar{f}(t) \text{ for } 0 \le t \le 1\}.$$

(ii) At the other extreme, suppose that X and A have the following property: For every x, y in X, $\bar{e}_x \ne e_y$. Thus $\tau(x) \notin X$ for each x in X. Hence Y is the disjoint union of X and $\tau(X)$.

As an example of this case, we may consider $A = C([0,1])$. We may note that if A is any complex function algebra, then A has property (ii), because if $x \neq y$, we can find f in A with $f(x) = 0, f(y) = 1$, and if $x = y$, we can take f as the constant function i.

Thus every complex function algebra on X can be regarded as a real function algebra on (Y,τ), where Y is a disjoint union of two copies of X and the involution τ sends a point in one copy of X to the corresponding point in the other copy. This justifies our earlier observation that the class of real function algebras includes that of complex function algebras.

Note that if A is a complex function algebra on a compact Hausdorff space X, then it is not possible to define any topological involution τ' on X in such a way that A becomes a real function algebra on (X,τ'), because the constant function i does not belong to $C(X,\tau')$ for any τ'.

(iii) In between the two extremes mentioned above, we may consider the following property: For every x,y in X with $x \neq y$, there is f in A such that $f(x) \neq \bar{f}(y)$. This happens, in particular, if Re A separates the points of X. Let

$$Z := \{x \in X : f(x) \text{ is a real for all } f \text{ in } A\}.$$

Then $\tau(x) = x$ for all x in Z and $\tau(x) \in Y\backslash X$ for all x in $X\backslash Z$. Thus Y is a disjoint union of X and $\tau(X\backslash Z)$. As an example of this case, we may consider

$$A := \{f \in C([0,1]) : f(\tfrac{1}{2}) \text{ is real}\}.$$

Given a real function algebra A, we will now show that there is a complex function algebra that can be viewed as a complexification of A (see Definition 1.1.6).

Theorem 1.3.20 Let X be a compact Hausdorff space and τ a topological involution on X. Let $\sigma(f) = \bar{f} \circ \tau$ be the algebra involution induced by τ on $C(X)$. Let A be a real subspace of $C(X,\tau)$ and define

$$B := \{f + ig : f,g \in A\}.$$

Then
(i) $\sigma(B) = B$ and $A = \{h \in B : \sigma(h) = h\} = B \cap C(X,\tau)$.
(ii) For f,g in $C(X,\tau)$,

$$\max\{\|f\|, \|g\|\} \leq \|f + ig\| \leq \|f\| + \|g\|.$$

(iii) B is uniformly closed (respectively, separates the points of X, contains 1, is an algebra) if and only if A is uniformly closed (respectively, separates the points of X, contains 1, is an algebra).
(iv) Let A be a real function algebra on (X,τ). For ϕ in Car(A), define

$$\alpha(\phi)(f + ig) := \phi(f) + i\phi(g)$$

for f, g in A. Then $\alpha(\phi) \in \text{Car}(B)$. The map α is a bijection between $\text{Car}(A)$ and $\text{Car}(B)$.

Proof. We shall use the properties of σ proved in Theorem 1.3.5.

(i) Let $h = f + ig \in B$, $f, g \in A$. Since σ is an algebra involution, $\sigma(h) = \sigma(f + ig) = f - ig \in B$. Also, if $\sigma(h) = h$, then $h = f \in A$. This proves (i).

(ii) Let $f, g \in C(X, \tau)$. Since σ is an isometry, $\|f + ig\| = \|\sigma(f + ig)\| = \|f - ig\|$. Hence $\|f\| \leq (\frac{1}{2})(\|f + ig\| + \|f - ig\|) = \|f + ig\|$. Similarly, $\|g\| \leq \|f + ig\|$. This proves (ii).

(iii) The "only if" parts are trivial. Let now A be uniformly closed. Let $f_n, g_n \in A$ and $f_n + ig_n$ be a sequence in B, converging to a function $h \in C(X)$. Then since σ is an isometry, $f_n - ig_n = \sigma(f_n + ig_n) \to \sigma(h)$ as $n \to \infty$. Hence $f_n \to (h + \sigma(h))/2$ and $g_n \to (h - \sigma(h))/2i$ as $n \to \infty$. Since A is uniformly closed, $(h + \sigma(h))/2$ and $(h - \sigma(h))/2i$ are in A. Hence $h = (h + \sigma(h))/2 + i(h - \sigma(h))/2i \in B$. This proves that B is uniformly closed.

Since $A \subset B$, if A contains 1, so does B, and if A separates the points of X, so does B. It is easy to see that if A is an algebra, then B is an algebra.

(iv) Let $\phi \in \text{Car}(A)$. It is straightforward to check that $\alpha(\phi) \in \text{Car}(B)$, $\alpha(\phi)|_A = \phi$, and for any $\psi \in \text{Car}(B)$, $\psi|_A \in \text{Car}(A)$ with $\alpha(\psi|_A) = \psi$. This proves (iv). $\qquad\square$

In view of property (ii) of Theorem 1.3.20, every h in B can be written *uniquely* as $h = f + ig$ for some f, g in A. Hence we can regard B as the *complexification of A* (Definition 1.1.6) whenever A is a real algebra. The supremum norm on B satisfies the properties of the norm p mentioned after Definition 1.1.6.

We may note that if X is a compact Hausdorff space and τ a topological involution on X, then $C(X)$ is the complexification of $C(X, \tau)$. In particular, $C(X)$ is the complexification of $C_\mathbb{R}(X)$. The disk algebra of Example 1.3.10 is the complexification of the real disk algebra. Further, the algebra B in Example 1.3.11 is the complexification of the algebra A in the same example. The same thing can be said about the algebras B and A of Example 1.3.12.

Corollary 1.3.21 Let X be a compact Hausdorff space and τ be a topological involution on X. Then

$$\text{Car}(C(X, \tau)) = \{e_x : x \in X\}.$$

Proof. Clearly, $e_x \in \text{Car}(C(X, \tau))$ for each x in X. Now, let $\phi \in \text{Car}(C(X, \tau))$. Then, by Theorem 1.3.20, $\alpha(\phi) \in \text{Car}(C(X))$. Hence, by Theorem 1.3.2, $\alpha(\phi) = e_x$ for some x in X. Hence $\phi = e_x$ on $C(X, \tau)$. $\qquad\square$

One can also give a direct proof of Corollary 1.3.21, which is similar to the proof of Theorem 1.3.2.

In some problems, we need to know when a given complex function algebra can be viewed as a complexification of a real function algebra. The following theorem gives a criterion.

Theorem 1.3.22 Let X be a compact Hausdorff space, τ a topological involution on X, σ the algebra involution on $C(X)$ given by $\sigma(f) = \bar{f} \circ \tau$, $f \in C(X)$, and B a complex function algebra on X. If $\sigma(B) = B$, then B can be regarded as a complexification of a real function algebra A on (X,τ) given by $A := \{h \in B : \sigma(h) = h\}$.

Proof. Let $h \in B$. Since $\sigma(B) = B$, we have $f := (h + \sigma(h))/2$ and $g := (h - \sigma(h))/2i$ belong to B. Further, since $\sigma(f) = f$ and $\sigma(g) = g$, $f, g \in A$. Also, $h = f + ig$. Using properties of σ, it is easy to prove that if $h = f_1 + ig_1$, with f_1, g_1 in A, then $f = f_1$ and $g = g_1$. Thus every element h of B can be uniquely expressed as $h = f + ig$ with f, g in A. Now it remains only to show that A is a real function algebra on (X,τ). It is straightforward to check that A is a real subalgebra of $C(X,\tau)$. Since σ is an isometry and A is the set of all fixed points of σ, A is uniformly closed. Also, $\sigma(1) = 1 \in A$. Now let $x, y \in X$ and $x \neq y$. Since B separates the points of X, there exists h in B such that $h(x) \neq h(y)$. Now $h = f + ig$ for some $f, g \in A$. Hence $f(x) \neq f(y)$ or $g(x) \neq g(y)$. Thus A separates the points of X. $\square$

Theorem 1.3.20 allows us to use the complexification technique mentioned after Definition 1.1.6 to deduce results for real function algebras from the corresponding results for complex function algebras. We shall, however, prefer to develop the theory of real function algebras intrinsically. Only when this seems infeasible shall we resort to the complexification technique, with a hope that someone will furnish intrinsic proofs.

Let X be a compact Hausdorff space, τ a topological involution on X, and A a real function algebra on (X,τ). The most natural question that one can ask about A is: When is A equal to the whole of $C(X,\tau)$? This question is the central theme of Chapter 2.

1.4 DEFORMATIONS OF A REAL FUNCTION ALGEBRA

In this section we study the effect of introducing a small change in the operation of multiplication in a real function algebra. This subject is yet to be developed to its full potential even in the case of complex function algebras. Our treatment is, therefore, necessarily cursory. The material in this section is not used subsequently in this book.

We first make precise the meaning of the terminology "small change." There are two essentially different ways of doing this.

Definition 1.4.1 Let A be a real (respectively, complex) Banach algebra and let the original multiplication in A be denoted by juxtaposition as usual. Let $\varepsilon > 0$. A binary operation $*$ defined on A is called an *algebraic ε-perturbation of A* if $(A,*)$ is a real (respectively, complex) algebra and

$$\|a * b - ab\| \leq \varepsilon \|a\| \|b\| \qquad \text{for all } a,b \text{ in } A.$$

The algebra $(A,*)$ is called an *algebraic ε-deformation of A*.

A is called *stable* (or *rigid*) if there exists $\varepsilon_0 > 0$ such that every algebraic ε-deformation of A is isomorphic to A for $0 \leq \varepsilon \leq \varepsilon_0$.

Definition 1.4.1 implies that the original multiplication and the new multiplication are close to each other as bilinear maps on A. Note that if $*$ is any continuous multiplication on A, then there exists $k > 0$ such that $\|a * b\| \leq k\|a\| \|b\|$ for all a,b in A and hence

$$\|a * b - ab\| \leq (k + 1)\|a\| \|b\| \qquad \text{for all } a,b \text{ in } A.$$

Thus $*$ is an algebraic $(k + 1)$-perturbation of A. We are, therefore, interested in ε-perturbations of A for small values of ε, particularly $\varepsilon < 1$ in many cases.

Perturbations of complex Banach algebras have been studied in recent years by Kadison and Kastler (1972), Philips (1973), Johnson (1977), and Raeburn and Taylor (1979). The investigations concerning the perturbations of complex function algebras were initiated by Rochberg (1979) and continued in several articles by Rochberg and Jarosz. The lecture notes by Jarosz (1985) contain a detailed account of all these results.

Definition 1.4.2 Let A be a real (respectively, complex) Banach algebra. Let $\varepsilon > 0$. A binary operation $*$ defined on A is called a *metric ε-perturbation of A* if there is a real (respectively, complex) Banach algebra $(A_1, \cdot)$ and a linear, continuous bijection $T : A \to A_1$ such that

$$a * b = T^{-1}(Ta \cdot Tb)$$

for all a,b in A and $\|T\| \|T^{-1}\| \leq 1 + \varepsilon$. A_1 is then called a *metric ε-deformation of A*.

It is easy to see that $(A,*)$ is a real (respectively, complex) algebra.

A natural question at this stage is whether an algebraic (or metric) deformation of a function algebra is a function algebra. The next theorem deals with this question for an algebraic perturbation. Since every function algebra is a uniform algebra (Definition 1.3.13), and vice versa (Theorem 1.3.17), it is enough to establish the results for uniform algebras.

Since $\|a * b\| \leq (1 + \varepsilon)\|a\| \|b\|$ for an ε-algebraic perturbation of A, the spectral radius of a in $(A, *)$ can be defined as in 1.1.12. Also, 1.1.13 and the result following it can be proved.

Theorem 1.4.3 Let A be a real or complex uniform algebra and $*$ be an algebraic ε-perturbation of A with $0 \leq \varepsilon < 1$. Let $N(f)$ denote the spectral radius of f in $(A, *)$. Then:
(i) $(1 - \varepsilon)\|f\| \leq N(f) \leq (1 + \varepsilon)\|f\|$ for all f in A.
(ii) If $*$ is commutative, then $(A, *, N)$ is a uniform algebra. Moreover, $*$ is a metric ε_1-perturbation of A, where $\varepsilon_1 = 2\varepsilon/(1 - \varepsilon)$.

Proof.
(i) Let $f \in A$. Then

$$\left| \|f * f\| - \|f^2\| \right| \leq \|f * f - f^2\| \leq \varepsilon \|f\|^2.$$

Since $\|f^2\| = \|f\|^2$, we get

$$(1 - \varepsilon)\|f\|^2 \leq \|f * f\| \leq (1 + \varepsilon)\|f\|^2.$$

By induction, we see that for all $n = 1, 2, \ldots,$

$$(1 - \varepsilon)^{2^n}(1 - \varepsilon)^{2^{n+1}}\|f\|^{2^n} \leq \underbrace{\|f * \cdots * f\|}_{2^n \text{ times}} \leq (1 + \varepsilon)^{2^n - 1}\|f\|^{2^n}.$$

Taking the 2^nth root of each term in the inequality above, letting $n \to \infty$, and using the spectral radius formula (Theorem 1.1.19), we obtain

$$(1 - \varepsilon)\|f\| \leq N(f) \leq (1 + \varepsilon)\|f\|.$$

(ii) (i) implies that $f = 0$ if and only if $N(f) = 0$. If $*$ is commutative, it follows from (iv) and (vii) of Theorem 1.2.9 that the spectral radius N is a norm on $(A, *)$. By Corollary 1.2.12, $N(f * f) = N(f)^2$ for all f in A. Now we show that $*$ has a unit. Define a map $T_1 : A \to A$ by $T_1(f) = 1 * f$ for f in A, and let I denote the identity map on A. For f in A,

$$\|1f - 1 * f\| \leq \varepsilon \|f\|, \quad \|1 * f\| \leq (1 + \varepsilon)\|f\|.$$

Hence T_1 is a bounded linear map on A and $\|I - T_1\| < 1$, so that T_1 is invertible [Lemma 1.1.14(i)]. Let $e = T_1^{-1}(1)$. Then $1 = T_1(e) = 1 * e$. Now for all f in A, $T_1(e * f) = 1 * (e * f) = (1 * e) * f = 1 * f = T_1(f)$. Since T_1 is injective, this implies that $e * f = f$ for all f in A. Hence e is a unit in $(A, *)$. Thus $(A, *)$ is a uniform algebra.

Next consider the map $T: A \to (A, *)$ given by $T(f) = f$ for all f in A. Then from (i), $\|T\| \le 1 + \varepsilon$ and $\|T^{-1}\| \le 1/(1 - \varepsilon)$. Thus $\|T\| \|T^{-1}\| \le (1 + \varepsilon)/(1 - \varepsilon)$. Since $(1 + \varepsilon)/(1 - \varepsilon) = 1 + 2\varepsilon/(1 - \varepsilon)$, T defines a metric ε_1-perturbation of A, where $\varepsilon_1 = 2\varepsilon/(1 - \varepsilon)$. $\qquad\square$

Notice that in (ii) of Theorem 1.4.3, we have assumed that $*$ is commutative. Jarosz (1985) showed that such an assumption is not necessary for complex uniform algebras. We shall now prove this result of Jarosz and use it along with the complexification technique to deduce a similar result for real function algebras. (See the comments at the end of Section 1.3.)

Lemma 1.4.4 Let X be a compact Hausdorff space, τ a topological involution on X, and A a real function algebra on (X, τ). Suppose that $*$ is an algebraic ε-perturbation of A. Let $B := \{f + ig : f, g \in A\}$. Define a binary operation, also denoted by $*$, on B as follows: For $h_1 = f_1 + ig_1, h_2 = f_2 + ig_2 \in B$ with $f_1, g_1, f_2, g_2 \in A$, let

$$h_1 * h_2 = f_1 * f_2 - g_1 * g_2 + i(f_1 * g_2 + g_1 * f_2).$$

Then $*$ is an algebraic 4ε-perturbation of B.

Proof. For h_1 and h_2 as above, we have

$$h_1 h_2 = f_1 f_2 - g_1 g_2 + i(f_1 g_2 + g_1 f_2).$$

Hence

$$
\begin{aligned}
\|h_1 * h_2 - h_1 h_2\| &\le \|f_1 * f_2 - f_1 f_2\| + \|g_1 * g_2 - g_1 g_2\| \\
&\quad + \|f_1 * g_2 - f_1 g_2\| + \|g_1 * f_2 - g_1 f_2\| \\
&\le \varepsilon(\|f_1\| \|f_2\| + \|g_1\| \|g_2\| + \|f_1\| \|g_2\| + \|g_1\| \|f_2\|) \\
&\le 4\varepsilon \|h_1\| \|h_2\|
\end{aligned}
$$

by Theorem 1.3.20(ii). Thus $*$ is an algebraic 4ε-perturbation of B. $\qquad\square$

Theorem 1.4.5 Let A be a real or complex Banach algebra with unit 1, and $*$ be an algebraic ε-perturbation of A. Suppose that one of the following conditions holds.
(i) A is a complex function algebra on a compact Hausdorff space X, and $\varepsilon < 1$.
(ii) A is a real function algebra on a compact Hausdorff space X with a topological involution τ, and $\varepsilon < 1/4$.
(iii) A is a real uniform algebra, there exists a positive constant k such that $\|f\|^2 \le k\|f^2 + g^2\|$ for all f, g in A, and $\varepsilon < 1/2k$.

Then $*$ is commutative, and $(A,*)$ is a uniform algebra under an equivalent norm.

Proof. In view of Theorem 1.4.3(ii), it is enough to show that $*$ is commutative in all the cases.

Let condition (i) hold and $N(f)$ denote the spectral radius of f in $(A,*)$. By Theorem 1.4.3(i), for all f in A,

$$\|f\|^2 \le \frac{N(f)^2}{(1-\varepsilon)^2} = \frac{N(f*f)}{(1-\varepsilon)^2}$$
$$= \frac{(1+\varepsilon)\|f*f\|}{(1-\varepsilon)^2}.$$

Now by Theorem 1.1.30(i), $*$ is commutative.

Next, let condition (ii) hold and $B := \{f + ig : f,g \in A\}$. Then B is a complex function algebra, and by Theorem 1.4.4, $*$ is an algebraic 4ε-perturbation of B. Since $4\varepsilon < 1$, $*$ is commutative by (i).

Finally, suppose that condition (iii) holds. Then for f,g in A,

$$\|f\|^2 \le k\|f^2 + g^2\|$$
$$\le k(\|f^2 - f*f\| + \|f*f + g*g\| + \|g*g - g^2\|)$$
$$\le k\|f*f + g*g\| + \varepsilon k(\|f\|^2 + \|g\|^2) \qquad (1)$$

and similarly for $\|g\|^2$. Hence

$$\|f\|^2 \le \|f\|^2 + \|g\|^2 \le \beta\|f*f + g*g\|, \qquad (2)$$

where $\beta = 2k/(1 - 2\varepsilon k) > 0$ as $2k\varepsilon < 1$. Hence $*$ is commutative by Theorem 1.1.30(ii). $\qquad \Box$

The proof of part (ii) of Theorem 1.4.5 is an illustration of employing the complexification technique to study the effect of perturbation on a property (in this case, commutativity) of a real function algebra. Some more examples of this technique can be found in Kulkarni and Arundhathi (1991–c).

We do not know whether the conclusion (ii) of Theorem 1.4.5 holds if $1/4 \le \varepsilon < 1$.

Note that if X is a compact Hausdorff space, then $\|f\|^2 \le \|f^2 + g^2\|$ for all f,g in $C_\mathbb{R}(X)$. Hence by part (iii) of Theorem 1.4.5, every algebraic ε-deformation of $C_\mathbb{R}(X)$ with $\varepsilon < 1/2$ is commutative.

Theorems 1.4.3 and 1.4.5 raise the following natural questions about metric perturbations and commutativity.

1. Is every small metric perturbation of a function algebra A a small algebraic perturbation? Precisely, is every metric ε-perturbation of A

with $0 \le \varepsilon < \varepsilon_0$ for some fixed ε_0, an algebraic $k(\varepsilon)$-perturbation of A, where $k(\varepsilon)$ is a function of ε such that $k(\varepsilon) \to 0$ as $\varepsilon \to 0$?

2. Is every small metric perturbation of a real or complex function algebra commutative? (Precise formulation as above can be given.)

3. Is every small metric deformation of a real or complex function algebra also a real or complex function algebra?

Notice that if question 1 has the affirmative answer, then so do questions 2 and 3, in view of Theorems 1.4.3 and 1.4.5. Jarosz (1985) has shown that this, in fact, is the case for complex function algebras. Now we show that the situation is rather different for real function algebras.

Example 1.4.6 Let $A = (\mathbb{R}^2, \|\cdot\|_\infty)$ with the coordinatewise multiplication. For a fixed t with $0 \le t \le \frac{1}{2}$, define $\rho_t \colon \mathbb{R}^2 \times \mathbb{R}^2 \to \mathbb{R}^2$ by

$$\rho_t((x_1,y_1), (x_2,y_2)) = (x_1 x_2 - t(x_1 - y_1)(x_2 - y_2),$$
$$y_1 y_2 - t(x_1 - y_1)(x_2 - y_2)).$$

A direct computation shows that $A_t := (\mathbb{R}^2, \rho_t, \|\cdot\|_\infty)$ is a commutative Banach algebra with unit $(1,1)$. Let $T \colon A \to A_t$ be the identity map. Since $\|T\| \|T^{-1}\| = 1 < 1 + \varepsilon$ for every $\varepsilon > 0$, ρ_t is a metric ε-perturbation of A for *every* $\varepsilon > 0$. If the answer to question 1 were affirmative, there would exist positive constants ε_0 and a function $k(\varepsilon)$ defined for $0 \le \varepsilon < \varepsilon_0$ such that $k(\varepsilon) \to 0$ as $\varepsilon \to 0$ and

$$\|\rho_t((x_1,y_1), (x_2,y_2)) - (x_1,y_1)(x_2,y_2)\|_\infty \le k(\varepsilon)\|(x_1,y_1)\|_\infty \|(x_2,y_2)\|_\infty$$

for all (x_1,y_1), $(x_2,y_2) \in \mathbb{R}^2$ and every ε with $0 \le \varepsilon < \varepsilon_0$. But this would mean that ρ_t coincides with the original multiplication, which is false. Also,

$$\rho_t((1,-1), (1,-1)) = (1 - 4t, 1 - 4t) = 0 \qquad \text{if } t = \tfrac{1}{4}.$$

Thus $A_{1/4}$, which is a metric ε-deformation of A for *every* $\varepsilon > 0$, is *not* a uniform algebra. This shows that the answers to questions 1 and 3 are negative for real function algebras.

The next example is related to question 2.

Example 1.4.7 Let $X = \{x_1,x_2,x_3,x_4\}$ be equipped with the discrete topology. Then X is a compact Hausdorff space. Define $\tau \colon X \to X$ by $\tau(x_1) = x_2$, $\tau(x_2) = x_1$, $\tau(x_3) = x_4$, and $\tau(x_4) = x_3$. Then τ is an involution on X. Denote by $\mathbb{H}$ the Banach algebra of real quaternions (see Example 1.1.4). Let $a = a_1 + ia_2$, $b = b_1 + ib_2$, and $f_{a,b}$ in $C(X,\tau)$ be defined by $f_{a,b}(x_1) = a$, $f_{a,b}(x_2) = \bar{a}$, $f_{a,b}(x_3) = b$, and $f_{a,b}(x_4) = \bar{b}$. Then define $T \colon C(X,\tau) \to \mathbb{H}$

by $T(f_{a,b}) = (1 + j)^{-1}(a_1 + ia_2 + jb_1 + kb_2)$. Note that the constant function $f_{1,1}$ is the unit in $C(X,\tau)$, and $T(f_{1,1}) = f_{1,1}$.

Suppose that $\|f_{a,b}\| \leq 1$. Then $|a| \leq 1$, $|b| \leq 1$ and

$$\|T(f_{a,b})\| = \|(1 + j)^{-1}\| \, \|a_1 + ia_2 + jb_1 + kb_2\|$$
$$= \frac{\sqrt{a_1^2 + a_2^2 + b_1^2 + b_2^2}}{\sqrt{2}} \leq 1.$$

Hence $\|T\| \leq 1$. Since $T(f_{1,1}) = f_{1,1}$, we see that $\|T\| = 1$. Also, for every $f_{a,b}$ in $C(X,\tau)$, we have

$$\|f_{a,b}\|^2 = \max\{a_1^2 + a_2^2, b_1^2 + b_2^2\} \leq a_1^2 + a_2^2 + b_1^2 + b_2^2$$
$$= 2\|T(f_{a,b})\|^2.$$

This shows that $\|T^{-1}\| \leq \sqrt{2}$. Since $T(f_{1,0}) = (1 + j)^{-1}$, we have $\|T^{-1}\| = \sqrt{2}$. Thus T induces a metric ε-perturbation on $C(X,\tau)$, where $\varepsilon = \sqrt{2} - 1$, given by

$$f_{a,b} * f_{c,d} = T^{-1}(T(f_{a,b})T(f_{c,d})) \qquad \text{for } f_{a,b}, f_{c,d} \text{ in } C(X,\tau).$$

It is easy to see that $*$ is noncommutative. For example, let $g = f_{i,-i}$ and $h = f_{-1,i}$. Then $Tg = i$, $Th = j$. Since $(Tg)(Th) \neq (Th)(Tg)$, we see that $g * h \neq h * g$.

Note that Example 1.4.7 does not settle question 2; that is, we do not know whether there exists $\varepsilon_0 > 0$ such that every metric ε-perturbation with $\varepsilon \leq \varepsilon_0$ is commutative. We only know that if such an ε_0 exists, it must be less than $\sqrt{2} - 1$.

2

When Does a Real Function Algebra Equal $C(X,\tau)$?

The history of questions of the type mentioned in the title of this chapter begins with the celebrated theorem of Weierstrass, which states that every continuous real-valued function f on a closed bounded interval $[a,b]$ is the uniform limit of a sequence of polynomials. In other words, every uniformly closed complex subalgebra of $C([a,b])$ that contains the constant function 1 and the function x (given by $x(t) = t$, $t \in [a,b]$) is the whole of $C([a,b])$. A major generalization of this theorem was achieved by M. H. Stone in 1937. This generalization, now well known as the Stone–Weierstrass theorem, can be stated as follows. Let A be a complex function algebra (Definition 1.3.8) on a compact Hausdorff space X. If A is closed under conjugation, that is, $\bar{f} \in A$ whenever $f \in A$, then $A = C(X)$.

Ever since its discovery, the Stone–Weierstrass theorem has permeated most of modern analysis. Also, the theorem has undergone a plethora of generalizations. In this chapter we discuss some of these generalizations and their analogs for real function algebras. In the first section we consider the theorems due to Bishop and Machado. In Section 2.2 we derive a few properties of antisymmetric sets and peak sets for a real function algebra A on (X,τ) and study the relationship between these sets and the corresponding sets for the complexification of A. Subsequent sections deal with the analogs for real function algebras of the Hoffman–Wermer theorem (Section 2.3), Wermer's theorem (Section 2.4), Bernard's theorem, and Sidney's theorem (Section 2.5). All these theorems impose some conditions on Re A, the space of all real parts of functions in A. These conditions imply that $A = C(X,\tau)$. The classical theorems about complex function algebras can be obtained as special cases. The classical theorem of Hoffman–Wermer is, in fact, proved in the process of proving its real analog.

2.1 ANALOG OF BISHOP–MACHADO THEOREM

A large number of generalizations of the Stone–Weierstrass theorem, which are important in the study of function algebras, arise from replacing the hypothesis that the algebra under consideration be closed under conjugation by some other hypothesis. We may note here that the disk algebra A of Example 1.3.10 is as far away from being conjugate-closed as possible. Since the functions in A are analytic in the disk, f must be constant, whenever both f and $\bar{f}$ are in A. This was used by Bishop to focus on the subsets of a compact Hausdorff space X, on which the functions in a complex function algebra B behave like analytic functions. These subsets were called antisymmetric sets. Bishop's theorem states that if a continuous function f coincides with a function in B on each antisymmetric set, then $f \in B$ (Bishop, 1961). The classical proof of Bishop's theorem, given by Glicksberg (1962), was based on an idea of de Branges (1959) and used many nontrivial tools from functional analysis like the Hahn–Banach theorem, the Krein–Milman theorem, and the Riesz representation theorem. This proof can be found in Rudin (1973) or Burckel (1972). Prolla (1971) extended this technique to the case of vector-valued functions. Machado (1977) formulated a quantitative version of Prolla's theorem and gave a completely elementary proof. A self-contained exposition of Machado's proof can be found in Burckel (1984). This proof was considerably simplified by Ransford (1984) using a technique employed by Brosowski and Deutsch (1981). Prolla (1988) has used these results to prove a generalization of the Bernstein approximation theorem. References to the literature on the Stone–Weierstrass theorem and its generalizations can be found in Semadeni (1971) and Burckel (1972).

We begin this section with Ransford's proof of Machado's theorem. First, we fix some notation. Let X be a compact Hausdorff topological space and E be a real or complex normed linear space. By $C_E(X)$ we denote the set of all continuous E-valued functions on X. When $E = \mathbb{R}$, we get the special case $C_\mathbb{R}(X)$. We write $C(X)$ instead of $C_\mathbb{C}(X)$. $C_E(X)$ is a linear space. For f in $C_E(X)$ and K a nonempty closed subset of X, we define

$$\|f\|_K := \sup\{\|f(x)\| : x \in K\}.$$

Note that since K is compact, the supremum above is attained at some point of K. We omit the subscript K, when $K = X$. Thus $\|f\| := \|f\|_X$. $C_E(X)$ is a normed linear space with this norm $\|\cdot\|$ and it is a Banach space if E is a Banach space. For a nonempty subset G of $C_E(X)$, a closed subset K of X and f in $C_E(X)$, we denote by $d_K(f, G)$ the distance between $f_{|K}$ and $G_{|K} := \{g_{|K} : g \in G\}$, defined as follows:

$$d_K(f, G) := \inf\{\|f - g\|_K : g \in G\}.$$

It is easy to see that if $K \subset L$, then $d_K(f, G) \le d_L(f, G)$. Further, if $G \subset H \subset C_E(X)$, then $d_K(f, H) \le d_K(f, G)$. As above, the subscript K will be dropped when $K = X$.

Definition 2.1.1 Let X be a compact Hausdorff space, A a real subalgebra of $C(X)$, and K a nonempty subset of X. K is called a *partially A-antisymmetric set* (or a *set of partial antisymmetry of A*) if for f in A, "$f_{|K}$ is real" implies that "$f_{|K}$ is a constant." A is said to be a *partially antisymmetric algebra* if X is a partially A-antisymmetric set. Note that K is partially A-antisymmetric if and only if Re g is constant whenever both g and $\bar{g}$ belong to $A_{|K}$.

Theorem 2.1.2 (Machado, 1977; Ransford, 1984) Let X be a compact Hausdorff space and E be a real or complex normed linear space. For f in $C_E(X)$ and a subset G of $C_E(X)$, there exists a minimal closed subset S of X such that

$$d(f, G) = d_S(f, G).$$

In case G is a real subspace of $C_E(X)$ and A is a real subalgebra of $C(X)$ containing 1 such that for all g in G and a in A, ag is in G, then S is partially A-antisymmetric.

Proof. Let $\mathscr{F}$ be the collection of all nonempty closed subsets T of X such that $d(f, G) = d_T(f, G)$. Clearly, $X \in \mathscr{F}$. Thus $\mathscr{F}$ is nonempty. $\mathscr{F}$ is partially ordered by the set inclusion. Let $\mathscr{G}$ be a totally ordered subfamily of $\mathscr{F}$ and let $K := \cap \{T : T \in \mathscr{G}\}$. Since each T in $\mathscr{G}$ is a nonempty compact set and $\mathscr{G}$ is totally ordered, K is nonempty. We now prove that $K \in \mathscr{F}$ by showing that $d_K(f, G) = d(f, G)$. Clearly, $d_K(f, G) \le d(f, G)$. Note that for each T in $\mathscr{F}$, we have $d(f, G) = d_T(f, G)$, so that for each g in G, the set

$$T_g := \{x \in T : \|f(x) - g(x)\| \ge d(f, G)\}$$

is a nonempty compact set and $\{T_g : T \in \mathscr{G}\}$ is totally ordered. Hence

$$\cap \{T_g : T \in \mathscr{G}\} = \{x \in K : \|f(x) - g(x)\| \ge d(f, G)\}$$

is nonempty. Thus $\|f - g\|_K \ge d(f, G)$ for all g in G; that is, $d_K(f, G) \ge$

$d(f,G)$. Now by Zorn's lemma, there is a minimal element S in $\mathcal{F}$. Suppose that G is a real subspace of $C_E(X)$ and A is a real subalgebra of $C(X)$ containing 1 such that for g in G and a in A, ag belongs to G.

We shall show that S is partially A-antisymmetric. Suppose that this is false. Then there exists h in A such that $h_{|S}$ is real but not constant. We may assume by taking a suitable linear combination of 1 and h that

$$\min_{x \in S} h(x) = 0 \quad \text{and} \quad \max_{x \in S} h(x) = 1.$$

Let

$$Y := \{x \in S : 0 \le h(x) \le \tfrac{2}{3}\},$$
$$Z := \{x \in S : \tfrac{1}{3} \le h(x) \le 1\}.$$

Then Y and Z are nonempty proper closed subsets of S with $S = Y \cup Z$. By the minimality of S, there exist g_Y, g_Z in G such that

$$\|f - g_Y\|_Y < d(f,G),$$
$$\|f - g_Z\|_Z < d(f,G). \tag{i}$$

For $n \ge 1$, let $h_n = (1 - h^n)^{2^n}$ and $g_n = h_n g_Y + (1 - h_n)g_Z$. We have $h_n \in A$, $0 \le h_n \le 1$ on S, and $g_n \in G$. At each x in S, $g_n(x)$ is a convex combination of $g_Y(x)$ and $g_Z(x)$. Hence from (i)

$$\|f - g_n\|_{Y \cap Z} < d(f,G). \tag{ii}$$

On $Y\backslash Z$, $0 \le h < \tfrac{1}{4}$. Hence

$$h_n = (1 - h^n)^{2^n} \ge 1 - 2^n h^n \ge 1 - (\tfrac{2}{3})^n.$$

On $Z\backslash Y$, $\tfrac{2}{3} < h \le 1$. Hence

$$h_n = (1 - h^n)^{2^n} \le (1 + h^n)^{-2^n} \le (2^n h^n)^{-1} \le (\tfrac{3}{4})^n.$$

Thus g_n converges uniformly to g_Y on $Y\backslash Z$ and to g_Z on $Z\backslash Y$. Combining this with (i) and (ii), we see that for large n,

$$\|f - g_n\|_S < d(f,G).$$

This contradicts the fact that $d(f,G) = d_S(f,G)$. □

For many interesting generalizations and applications of Theorem 2.1.2, we refer the reader to Prolla (1988). We mention only one simple application, which we shall use later.

Corollary 2.1.3 (Prolla, 1988) Let A be a subalgebra of $C_{\mathbb{R}}(X)$ containing the constant function 1 and $f \in C_{\mathbb{R}}(X)$. Then f belongs to the closure of A if and only if f satisfies the following condition: For every

x, y in X such that $f(x) \neq f(y)$, there exists g in A such that $g(x) \neq g(y)$.

$$(*)$$

Proof. The condition (*) is obviously necessary. To establish the sufficiency, we may note that by Theorem 2.1.2, there exists a partially A-antisymmetric set K such that $d_K(f, A) = d(f, A)$.

Now since every function in A is real-valued, (*) implies that $f|_K$ is constant, say α. Then $\alpha \cdot 1 \in A$ and $\|f - \alpha \cdot 1\|_K = 0$. Thus $d(f, A) = d_K(f, A) = 0$, so that f belongs to the closure of A. $\qquad \square$

Corollary 2.1.4 Let X be a compact Hausdorff space and τ a topological involution on X. Let

$$C_{\mathbb{R}}(X, \tau) := \{f \in C_{\mathbb{R}}(X) : f \circ \tau = f\}.$$

Let A be a subalgebra of $C_{\mathbb{R}}(X, \tau)$ containing 1 and separating the points of X/τ. Then A is uniformly dense in $C_{\mathbb{R}}(X, \tau)$.

Proof. Consider $f \in C_{\mathbb{R}}(X, \tau)$, which is not constant on X. Let $x, y \in X$ be such that $f(x) \neq f(y)$. Then $y \neq x$ and $y \neq \tau(x)$. Since A separates the points of X/τ, there is $g \in A$ such that $g(x) \neq g(y)$. Thus condition (*) of Corollary 2.1.3 is satisfied. Hence f belongs to the closure of A. Since this is true for every nonconstant function f in $C_{\mathbb{R}}(X, \tau)$, and since $1 \in A$, it follows that the closure of A equals $C_{\mathbb{R}}(X, \tau)$. $\qquad \square$

If τ is the identity map on X, then Corollary 2.1.4 says that if A is a uniformly closed subalgebra of $C_{\mathbb{R}}(X)$ containing 1 and separating the points of X, then $A = C_{\mathbb{R}}(X)$. This is the classical Stone–Weierstrass theorem for $C_{\mathbb{R}}(X)$.

Theorem 2.1.5 Let X be a compact Hausdorff space and A be a real subalgebra of $C(X)$.
(i) If K and H are partially A-antisymmetric sets with nonempty intersection, then $K \cup H$ is partially A-antisymmetric.
(ii) Let $\{K_\alpha : \alpha \in \Lambda\}$ be a family of partially A-antisymmetric sets such that $\cap \{K_\alpha : \alpha \in \Lambda\}$ is nonempty. Then $K = \cup \{K_\alpha : \alpha \in \Lambda\}$ is partially A-antisymmetric.
(iii) Every partially A-antisymmetric set is contained in a maximal partially A-antisymmetric set.
(iv) Each x in X lies in a maximal partially A-antisymmetric set.
(v) The closure of a partially A-antisymmetric set is also partially A-antisymmetric.
(vi) Every maximal partially A-antisymmetric set is closed.
(vii) Distinct maximal partially A-antisymmetric sets are disjoint.

Proof.

(i) Let $x \in K \cap H$ and $f \in A$ be such that $f_{|K \cup H}$ is real. Then since K and H are partially A-antisymmetric, $f_{|K} \equiv f(x) \equiv f_{|H}$. Hence $f_{|K \cup H} \equiv f(x)$.

(ii) This can be proved in the same way as above. Notice that (i) is a special case of (ii).

(iii) Let K be a partially A-antisymmetric set. Let H be the union of all partially A-antisymmetric sets containing K. Then from (ii), H is partially A-antisymmetric. Clearly, if any partially A-antisymmetric set G contains H, then $G \supset K$ and hence $G \subset H$; that is, $G = H$. Thus H is a maximal partially A-antisymmetric set containing K.

(iv) This follows from (iii) because for each x in X, $\{x\}$ is a partially A-antisymmetric set.

(v) Let K be a partially A-antisymmetric set and H be its closure. Suppose that $f \in A$ is real on H. Then f is real on K and hence constant on K. By the continuity of f, $f_{|H}$ is constant.

(vi) Let K be a maximal partially A-antisymmetric set and H be its closure. By (v), H is partially A-antisymmetric. Hence by the maximality of K, $H \subset K$ so that K equals its closure H.

(vii) Let H and K be maximal partially A-antisymmetric sets. If $H \cap K$ is nonempty, then by (i) $H \cup K$ is partially A-antisymmetric. Hence by the maximality of H and K, we have $H \cup K \subset K$ and $H \cup K \subset H$. Thus $H = K$. $\qquad\square$

Corollary 2.1.6 (Bishop's theorem) Let X be a compact Hausdorff space and A be a uniformly closed real subalgebra of $C(X)$ containing 1. Let $f \in C(X)$. If $f_{|K} \in A_{|K} := \{f|_K : f \in A\}$ for every maximal partially A-antisymmetric set K, then $f \in A$.

Proof. By Theorem 2.1.2, there exists a closed partially A-antisymmetric set S such that $d(f,A) = d_S(f,A)$. By Theorem 2.1.5(iii), S is contained in a maximal partially A-antisymmetric set K. Now, as $S \subset K$ and $f_{|K} \in A_{|K}$,

$$d(f,A) = d_S(f,A) \leq d_K(f,A) = 0.$$

Since A is closed, we see that $f \in A$. $\qquad\square$

Remark 2.1.7 Let us apply Bishop's result to the situation when A is, in fact, a *complex* subalgebra of $C(X)$ containing 1. Then for every $f \in C(X)$ and every $x \in X$, $f_{|\{x\}} \in A_{|\{x\}}$. Thus if every maximal partially A-antisymmetric set is a singleton $\{x\}$, then for every $f \in C(X)$, the hypothesis in Bishop's theorem is satisfied, so that $A = C(X)$. This is, for example, the case when A separates the points of X and $\bar{f} \in A$ whenever $f \in$

A. For let *K* be a partially *A*-antisymmetric set. If possible, let $x, y \in K$, $x \neq y$. Find $g \in A$ with $g(x) \neq g(y)$. Then either Re $g(x) \neq$ Re $g(y)$ or Im $g(x) \neq$ Im $g(y)$. Since Re $g = (g + \bar{g})/2$ as well as Im $g = (g - \bar{g})/2i$ belong to *A* and are real-valued, we obtain a contradiction to the partial *A*-antisymmetry of *K*. We have thus proved the *Stone–Weierstrass theorem*: If a complex function algebra *A* on *X* is conjugate closed, then $A = C(X)$.

We would like to obtain similar results for real function algebras, but since multiplication by *i* is not possible, the arguments above will no longer apply. We therefore strengthen the concept of a partially *A*-antisymmetric set as follows.

Definition 2.1.8 Let *X* be a compact Hausdorff space, *A* a real subalgebra of $C(X)$, and *K* a nonempty subset of *X*. Then *K* is called an *A-antisymmetric set* if for *f* in *A*, either "Im$(f_{|K}) = 0$" or "Re$(f_{|K}) = 0$" implies that $f_{|K}$ is constant. Also, *A* is said to be an *antisymmetric algebra* if *X* is an *A*-antisymmetric set.

Note that every singleton set $\{x\}$, $x \in X$, is trivially *A*-antisymmetric. It is clear that every *A*-antisymmetric set is partially *A*-antisymmetric, while if *A* contains *i* or if $A \subset C_{\mathbb{R}}(X)$, then the converse also holds. In general, however, a partially *A*-antisymmetric set may not be *A*-antisymmetric, as the following example shows. Let

$$A := \{f \in C([0,1]) : f(t) = \bar{f}(1 - t) \text{ for all } t \in [0,1]\}.$$

Then $K = \{\frac{1}{4}, \frac{3}{4}\}$ is clearly a partially *A*-antisymmetric set, but not an *A*-antisymmetric set because the function $f(t) = i(2t - 1)$, $t \in [0,1]$, belongs to *A*, $f(\frac{1}{4}) = -i/2$ and $f(\frac{3}{4}) = i/2$.

We may note here the reason for the nomenclature "antisymmetric." For an *A*-antisymmetric set *K*, the restriction algebra $A_{|K}$ is very far from being symmetric, that is, conjugate-closed. In fact, *K* is *A*-antisymmetric if and only if *g* is constant on *K* whenever *g* and $\bar{g}$ both belong to $A_{|K}$. Similarly, the nomenclature "partially antisymmetric" introduced earlier can be explained.

Remark 2.1.9 It is easy to check that the statement and the proof of Theorem 2.1.5 remain valid if we replace the words "partially *A*-antisymmetric" by "*A*-antisymmetric" everywhere. One consequence of this fact is that the space *X* is partitioned into maximal *A*-antisymmetric sets. This partition is called the *Bishop decomposition of X by A*. For a related concept of Shilov decomposition, we refer the reader to Shilov (1951).

We now consider an extension of Theorem 2.1.2 of Machado to the case of a real function algebra. First we show that in this situation the partially antisymmetric and antisymmetric sets have additional properties.

Theorem 2.1.10 Let X be a compact Hausdorff space and τ a topological involution on X. Let A be a (real) subalgebra of $C(X,\tau)$.
(a) Let K be a partially A-antisymmetric subset of X. Then we have the following results:

(i) $\tau(K)$ and $K \cup \tau(K)$ are partially A-antisymmetric.
(ii) If K is a maximal partially A-antisymmetric set, then $\tau(K) = K$.
(iii) If $\tau(K) = K$, then either K is A-antisymmetric or K is the disjoint union of S and $\tau(S)$, where S is A-antisymmetric.

(b) Let K be an A-antisymmetric subset of X. Then we have the following results:

(i) $\tau(K)$ is A-antisymmetric, while $K \cup \tau(K)$ is A-antisymmetric if and only if $f_{|K} \equiv 0$ whenever $f \in A$ and $\mathrm{Re}(f_{|K}) \equiv 0$. In particular, if $K \cup \tau(K)$ is not A-antisymmetric, then $K \cap \tau(K) = \varnothing$ and there is h in A such that $h_{|K} \equiv i$ and $h_{|\tau(K)} \equiv -i$.
(ii) If K is a maximal A-antisymmetric set, so is $\tau(K)$ and either $\tau(K) = K$ or $K \cap \tau(K) = \varnothing$. In the former case, if the real part of f in A vanishes on K, then f vanishes on K. In the latter case, there is h in A such that $h \equiv i$ on K and $h \equiv -i$ on $\tau(K)$. In both cases, $K \cup \tau(K)$ is a maximal partially A-antisymmetric set.

Proof.
    (a)(i): Let $f \in A$ and $f_{|\tau(K)}$ be real. Since $f(x) = \bar{f}(\tau(x))$ for all x, $f_{|K}$ is real and hence a constant, say c. But then $f_{|\tau(K)} = \bar{c}$. Thus $\tau(K)$ is partially A-antisymmetric.
    Now let $f \in A$ be real on $K \cup \tau(K)$. Then $f_{|K}$ is a constant, say c. But then $f_{|\tau(K)} = \bar{c} = c$. Thus f is a constant on $K \cup \tau(K)$. Hence $K \cup \tau(K)$ is partially A-antisymmetric.
    (a)(ii): By (a)(i), $K \cup \tau(K)$ is partially A-antisymmetric. By the maximality of K, $K \cup \tau(K) \subset K$. Hence $\tau(K) = K$.
    (a)(iii): If K is not A-antisymmetric, there is h in A such that $\mathrm{Re}(h_{|K}) = 0$, but $h_{|K} \neq$ constant. Then $h^2_{|K}$ is real and hence constant. Since this must be a nonzero constant, we can assume that $h^2_{|K} = -1$. Thus $h(x) = \pm i$ for all x in K. Let

$$S := \{x \in K : h(x) = i\} \text{ and } T := \{x \in K : h(x) = -i\}.$$

Clearly, $K = S \cup T$. Since h is nonconstant, both S and T are nonempty.

Also, S and T are disjoint. Now $\tau(S) \subset \tau(K) = K$. Further, for $x \in S$, $\tau(x) \in \tau(S) \subset K$ and $h(\tau(x)) = \bar{h}(x) = -i$. Hence $\tau(x) \in T$. Thus $\tau(S) \subset T$. Similarly, we can show that $\tau(T) \subset S$, so that $T = \tau(S)$, and K is a disjoint union of S and $\tau(S)$. It remains only to prove that S is A-antisymmetric. Let $f \in A$ be real on S. Then f is real on $\tau(S)$. Hence f is real on $S \cup \tau(S) = K$. Since K is partially A-antisymmetric, f is constant on K and hence on S. This proves that S is partially A-antisymmetric. Now let $g \in A$ and $\mathrm{Re}(g_{|S}) \equiv 0$. Then $(hg)_{|S} = (ig)_{|S}$ is real on S. Hence $(ig)_{|S} \equiv$ constant. This implies that $g_{|S}$ is a constant and S is A-antisymmetric.

(b)(i): By (a)(i), $\tau(K)$ is partially A-antisymmetric. Now let $f \in A$ and $\mathrm{Re}(f_{|\tau(K)}) \equiv 0$. Then for x in K, $\mathrm{Re}\, f(x) = \mathrm{Re}\, \bar{f}(\tau(x)) = \mathrm{Re}\, f(\tau(x)) = 0$. Hence $\mathrm{Re}(f_{|K}) \equiv 0$. As K is A-antisymmetric, it follows that f is constant on K, say c. But then, $f|_{\tau(K)} \equiv \bar{c}$. This shows that $\tau(K)$ is A-antisymmetric.

Next let $K \cup \tau(K)$ be A-antisymmetric. Consider f in A with $\mathrm{Re}(f_{|K}) \equiv 0$. Then it follows that $\mathrm{Re}(f_{|K \cup \tau(K)}) \equiv 0$, and hence $f_{|K \cup \tau(K)}$ is a constant, say, $i\alpha$ for some real α. Now for x in K, $i\alpha = f(x) = \bar{f}(\tau(x)) = -i\alpha$. Hence $\alpha = 0$; that is, $f_{|K} \equiv 0$. Conversely, assume that $f_{|K} \equiv 0$ whenever $f \in A$ and $\mathrm{Re}(f_{|K}) \equiv 0$. Since K is A-antisymmetric, it is partially A-antisymmetric and by (a)(i) above, $K \cup \tau(K)$ is partially A-antisymmetric. Now, if $f \in A$ and $\mathrm{Re}(f_{|K \cup \tau(K)}) \equiv 0$, then $\mathrm{Re}(f_{|K}) \equiv 0$. Hence by the assumption, $f_{|K} \equiv 0$, so that $f_{|K \cup \tau(K)} \equiv 0$. Thus $K \cup \tau(K)$ is A-antisymmetric.

Now suppose that $K \cup \tau(K)$ is not A-antisymmetric. Then by what we have just now proved, there is f in A such that $\mathrm{Re}(f_{|K}) \equiv 0$, but $f_{|K} \neq 0$. Since K is A-antisymmetric, $f_{|K}$ is a constant, say, $i\alpha$ for some real nonzero α. If we let $h = f/\alpha$, then $h_{|K} \equiv i$ and $h_{|\tau(K)} \equiv -i$. This also shows that K and $\tau(K)$ are disjoint.

(b)(ii): Let K be a maximal A-antisymmetric set. That $\tau(K)$ is a maximal A-antisymmetric set follows easily from (b)(i). Hence by Theorem 2.1.5 and Remark 2.1.9, either $K = \tau(K)$ or $K \cap \tau(K) = \varnothing$. In the former case, $K \cup \tau(K) = K$ is A-antisymmetric and hence by (b)(i), f vanishes on K whenever f belongs to A and $\mathrm{Re}\, f$ vanishes on K. In the latter case, $K \cup \tau(K)$ contains K properly and hence cannot be A-antisymmetric. Again the conclusion follows from (b)(i).

Next we prove that $K \cup \tau(K)$ is a maximal partially A-antisymmetric set. $K \cup \tau(K)$ is partially A-antisymmetric by (a)(i) and is hence contained in a maximal partially A-antisymmetric set L by Theorem 2.1.5. Now, by (a)(ii) and (a)(iii), L is either A-antisymmetric, or is a disjoint union of S and $\tau(S)$, where S, $\tau(S)$ are A-antisymmetric sets. If L is A-antisymmetric, then by the maximality of K, $L = K \subset K \cup \tau(K) \subset L$. Thus $L = K = K \cup \tau(K)$. Now suppose that $L = S \cup \tau(S)$, where S,

$\tau(S)$ are A-antisymmetric, and $S \cap \tau(S) = \varnothing$. Since $S \cup \tau(S)$ is not A-antisymmetric, by (b)(i) there is h in A such that $h_{|S} \equiv i$ and $h_{|\tau(S)} \equiv -i$. As Re $h_{|K} = 0$, $h_{|K}$ is a constant and since $K \subset K \cup \tau(K) \subset L = S \cup \tau(S)$, this constant must be either i or $-i$. Hence $K \subset S$ or $K \subset \tau(S)$. Again by the maximality of K, $K = S$ or $K = \tau(S)$. Hence $K \cup \tau(K) = S \cup \tau(S) = L$. Thus $K \cup \tau(K) = L$ is a maximal partially A-antisymmetric set. $\qquad \square$

As an example, let $A := \{f \in C([0,1]) : f(t) = \bar{f}(1 - t)$ for all t in $[0,1]\}$. Then A is a real function algebra on $([0,1], \tau)$, where $\tau(t) = 1 - t$, $t \in [0,1]$. It is easy to check that every singleton subset of $[0,1]$ is a maximal A-antisymmetric set. If $K = \{\frac{1}{2}\}$, then $K = \tau(K)$. Note that each f in A is real on K. If we take $L = \{0\}$, then $\tau(L) = \{1\}$, so that $L \cap \tau(L) = \varnothing$. The function $f(t) = i(1 - 2t)$, $t \in [0,1]$, belongs to A, takes the value i on L and $-i$ on $\tau(L)$.

We are now in a position to prove an analog of the theorem of Machado. We shall use the following notation.

Let X be a compact Hausdorff space, τ a topological involution on X, E a real normed linear space, and $*$ a linear involution on E. We define

$$C_E^*(X,\tau) := \{f \in C_E(X) : f(\tau(x)) = (f(x))^* \quad \text{for all } x \text{ in } X\}.$$

Thus $C(X,\tau)$ is a particular case of $C_E^*(X,\tau)$ when $E = \mathbb{C}$ and $*$ denotes conjugation. If $E = \mathbb{R}$ and $*$ is the identity map, we obtain the space

$$C_\mathbb{R}(X,\tau) := C_\mathbb{R}(X) \cap C(X,\tau) = \{f \in C_\mathbb{R}(X) : f \circ \tau = f\}.$$

Remark 2.1.11 Let X, τ, E, and $*$ be as above. If the linear involution $*$ is also an isometry (i.e., $\|a^*\| = \|a\|$ for all a in E), then for a subset K of X and f in $C_E^*(X,\tau)$, we have

$$\begin{aligned}
\|f\|_K &:= \sup\{\|f(x)\| : x \in K\} \\
&= \sup\{\|(f(x))^*\| : x \in K\} \\
&= \sup\{\|f(\tau(x))\| : x \in K\} \\
&=: \|f\|_{\tau(K)}.
\end{aligned}$$

Hence $\|f\|_K = \|f\|_{\tau(K)} = \|f\|_{K \cup \tau(K)}$.

Now, if G is any subset of $C_E^*(X,\tau)$, then

$$\begin{aligned}
d_K(f,G) &:= \inf\{\|f - g\|_K : g \in G\} \\
&= \inf\{\|f - g\|_{\tau(K)} : g \in G\} \\
&=: d_{\tau(K)}(f,G).
\end{aligned}$$

Thus $d_K(f,G) = d_{\tau(K)}(f,G) = d_{K \cup \tau(K)}(f,G)$.

Theorem 2.1.12 (Analog of Machado's theorem) Let X be a compact Hausdorff space, τ a topological involution on X, E a real normed linear space, and $*$ a linear involution on E which is an isometry. Let A be a subalgebra of $C(X,\tau)$ containing 1 and G a (real) subspace of $C_E^*(X,\tau)$ such that if $h \in A$ and $g \in G$, then $hg \in G$. Then for every f in $C_E^*(X,\tau)$, there exists a closed A-antisymmetric set S such that $d(f,G) = d_S(f,G)$.

Proof. By Machado's theorem (Theorem 2.1.2), there exists a closed partially A-antisymmetric set K, such that $d(f,G) = d_K(f,G)$. Let $T = K \cup \tau(K)$. Then by Theorem 2.1.10(a)(i), T is partially A-antisymmetric and by Remark 2.1.11, $d_K(f,G) = d_T(f,G)$. Since $\tau(T) = T$, it follows by Theorem 2.1.10(a)(iii) that either T is A-antisymmetric or $T = S \cup \tau(S)$, where S is A-antisymmetric. Since $d_T(f,g) = d_S(f,g)$, we are through. □

Corollary 2.1.13 (Analog of Bishop's theorem) Let X be a compact Hausdorff space, τ a topological involution on X, and A a uniformly closed subalgebra of $C(X,\tau)$ containing 1. If $f \in C(X,\tau)$ is such that $f_{|K} \in A_{|K}$ for every maximal A-antisymmetric set K in X, then $f \in A$. In particular, if every maximal A-antisymmetric set is a singleton, then $A = C(X,\tau)$.

Proof. We give two proofs for the first part. The first proof is similar to that of Corollary 2.1.6. By Theorem 2.1.12, there is an A-antisymmetric set S such that $d_S(f,A) = d(f,A)$. S is contained in a maximal A-antisymmetric set K by Remark 2.1.9 and Theorem 2.1.5(iii). Now, $d(f,A) = d_S(f,A) \leq d_K(f,A) = 0$, since $f_{|K} \in A_{|K}$. As A is closed, we see that $f \in A$.

The second proof uses Corollary 2.1.6. We first note that $f_{|S} \in A_{|S}$ for every A-antisymmetric set S. S is contained in a maximal A-antisymmetric set T by Remark 2.1.9. Now by the hypothesis, there is g in A such that $g = f$ on T. Hence $g = f$ on S.

Now let K be a maximal partially A-antisymmetric set. Then $K = \tau(K)$ by Theorem 2.1.10(a)(ii). Hence by Theorem 2.1.10(a)(iii), either K is A-antisymmetric or K is a disjoint union of S and $\tau(S)$, where S is A-antisymmetric. If K is A-antisymmetric, $f_{|K} \in A_{|K}$. Now suppose that K is a disjoint union of S and $\tau(S)$, where S is A-antisymmetric. Then as noted above, $f_{|S} \in A_{|S}$; that is, there is g in A such that $g(x) = f(x)$ for all x in S. But then $g(\tau(x)) = \bar{g}(x) = \bar{f}(x) = f(\tau(x))$ for all x in S. Hence $g(x) = f(x)$ for all x in $S \cup \tau(S) = K$. Thus we have proved that $f_{|K} \in A_{|K}$ in both the cases. Hence $f \in A$ by Corollary 2.1.6.

Now assume that every maximal A-antisymmetric set is a singleton. Let $f \in C(X,\tau)$, $x \in X$, $K = \{x\}$, and $f(x) = \alpha + i\beta$. If $x = \tau(x)$, then $f(x) = \alpha$ is real and $f_{|K} \in A_{|K}$ as the constant function α is in A. If $x \neq \tau(x)$, by

Theorem 2.1.10(b)(i), there is h in A with $h(x) = i$ and $h(\tau(x)) = -i$. Then $f_{|K} \in A_{|K}$ since $\alpha + \beta h \in A$. Since this is true for every $K = \{x\}, x \in X$, we see that $f \in A$. This implies that $A = C(X,\tau)$.

Corollary 2.1.14 (Analog of the Stone–Weierstrass theorem) Let X be a compact Hausdorff space, τ a topological involution on X, and A a real function algebra on (X,τ). If $\bar{f} \in A$ whenever $f \in A$, then $A = C(X,\tau)$.

Proof. We shall prove that every A-antisymmetric set is a singleton. Then the conclusion follows from Corollary 2.1.13. Let, if possible, some A-antisymmetric set K contain two distinct points x and y. There exists $f = u + iv$ in A such that $f(x) \neq f(y)$. Hence either $u(x) \neq u(y)$ or $iv(x) \neq iv(y)$. But $u = (f + \bar{f})/2$ and $iv = (f - \bar{f})/2$ belong to A, where u is real on K and $\mathrm{Re}(iv_{|K}) \equiv 0$. This contradicts the antisymmetry of K. □

Corollary 2.1.6 is a slight generalization of the classical theorem of Bishop. Theorem 2.1.12 and Corollaries 2.1.13 and 2.1.14 were proved in Kulkarni and Srinivasan (1987). Kulkarni and Limaye (1981b) give a different proof of Corollary 2.1.14.

Theorem 2.1.15 [Compare Grzesiak (1989)] Let A be a complex function algebra on X or a real function algebra on (X,τ). Then every finite A-antisymmetric set is a singleton. Further, if every A-antisymmetric set is finite, then $A = C(X)$ if A is a complex function algebra and $A = C(X,\tau)$ if A is a real function algebra.

Proof. Let $K = \{x_1,\ldots,x_n\}$ be a finite subset of X, where the x_j's are distinct and $n \geq 2$. We show that K cannot be A-antisymmetric. Let A be a complex function algebra on X. Then A separates the points of X, and for each $j = 2,\ldots,n$ there is a function $f_j \in A$ such that $f_j(x_1) = 1, f_j(x_j) = 0$. Then $f := f_2 \cdots f_n$ belongs to A, is real, but is not constant on K. Thus K cannot be A-antisymmetric. Next, let A be a real function algebra on (X,τ). If $\tau(x_1) \neq x_j$ for every $j = 2,\ldots,n$, then the proof above works in view of Lemma 1.3.9(ii) and shows that K is not A-antisymmetric. If $\tau(x_1) = x_m$ for some m, $2 \leq m \leq n$, then by Lemma 1.3.9(i), we can find $f_m \in A$ such that $f_m(x_1) = i, f_m(x_m) = -i$. Also, by Lemma 1.3.9(ii) we can find $f_j \in A$ for $j = 2,\ldots,n, j \neq m$, such that $f_j(x_1) = 1 = f_j(x_m), f_j(x_j) = 0$. Then $f = f_2 \cdots f_n$ belongs to A, $f(x_1) = i, f(x_m) = -i$, and $f(x_j) = 0$ for $j = 2,\ldots,n, j \neq m$. Since $\mathrm{Re}(f_{|K}) = 0$ but f is not constant on K, we see that K cannot be A-antisymmetric.

Finally, if every A-antisymmetric set is finite, then, in particular, every maximal A-antisymmetric set is finite and hence a singleton. Corollary 2.1.6 (respectively, Corollary 2.1.13) shows that $A = C(X)$ [respectively, $C(X,\tau)$] if A is a complex (respectively, real) function algebra.

□

Remark 2.1.16 Let A be a real function algebra on a compact Hausdorff space X with a topological involution τ. We have described in Remark 2.1.9 the Bishop decomposition of X consisting of all maximal A-antisymmetric sets. A decomposition or a cover $\mathfrak{S}$ of X by closed sets is said to *localize A* if whenever f is in $C(X,\tau)$ and $f_{|S}$ belongs to the uniform closure of $A_{|S}$ for each S in $\mathfrak{S}$, f belongs to A. Corollary 2.1.13 says that the bishop decomposition localizes A. Other decompositions/covers that localize A have been studied by Mehta, Mehta, and Vasavada (1990) and by Grzesiak (1991). We mention only two of them; information about the other decompositions/covers and references to similar investigations in the context of a complex function algebra can be found in these two references.

Let $\mathfrak{S}_1$ and $\mathfrak{S}_2$ be decompositions/covers of X. We say that $\mathfrak{S}_1$ is *finer* if every S_1 in $\mathfrak{S}_1$ is contained in some S_2 in $\mathfrak{S}_2$. For example, the Bishop decomposition is finer than the decomposition consisting of maximal partially A-antisymmetric sets. It is easy to see that if $\mathfrak{S}_1$ is finer than $\mathfrak{S}_2$ and $\mathfrak{S}_1$ localizes A, then $\mathfrak{S}_2$ localizes A.

The *Shilov decomposition of A*, as defined by Mehta and Vasavada, consists of the maximal sets of constancy of

$$A_R + A_I = \{f + g : f \in A, g \in A, \text{Im } f = 0 = \text{Re } g\}.$$

It is straightforward to check that the Bishop decomposition of A is finer than the Shilov decomposition of A since every function in $A_R + A_I$ is constant on an A-antisymmetric set. Hence the Shilov decomposition localizes A. This can be considered as a corollary of the analog of Bishop's theorem (Corollary 2.1.13).

We now introduce the concept of an annihilating measure. By $b(A\perp)$, we denote the set of all regular Borel measures μ on X such that the total variation of μ is at most 1, $\overline{\mu \circ \tau} = \mu$, and $\int_X f\, d\mu = 0$ for all f in A; μ is called an *extreme annihilating measure* if μ is an extreme point of $b(A\perp)$. Grzesiak (1991) has proved that the support of an extreme annihilating measure is an A-antisymmetric set. [In fact, a similar result was a crucial step in the classical proof of Bishop's theorem for complex function algebras; see Burckel (1972).] He then shows that the cover consisting of the supports of extreme annihilating measures along with the singleton subsets of X localizes A and is finer than the Bishop decomposition of A. Thus this result strengthens the analog of Bishop's theorem (Corollary 2.1.13).

The annihilating measures are also used by Hwang to discuss the following question: What conditions on a complex function algebra B ensure that $B \cap C(X,\tau)$ is a real function algebra on (X,τ)? We have given one such condition in Theorem 1.3.22, namely, $\sigma(B) = B$. Hwang (1990) has given the following sufficient condition: If λ and μ are regular Borel measures such that $\int_X f\,d\lambda = 0$ for all f in B and $\overline{\mu \circ \tau} = -\mu$, then λ and μ are mutually singular; that is, their supports are disjoint. He then asks whether there is any relationship between this condition and the condition $\sigma(B) = B$.

We shall now study the relationship between the antisymmetric sets of a real function algebra and those of its complexification. We shall use the properties of the complexification given in Theorem 1.3.20 and of the map σ given in Theorem 1.3.5.

Theorem 2.1.17 Let X be a compact Hausdorff space, τ be a topological involution on X, A be a real function algebra on (X,τ), and $B := \{f + ig, f,g \in A\}$ be the complexification of A.
(i) If K is a B-antisymmetric set, then K is also A-antisymmetric.
(ii) If K is a B-antisymmetric set, then $\tau(K)$ is also B-antisymmetric.
(iii) If K is an A-antisymmetric set such that $\tau(K) = K$, then K is B-antisymmetric.
(iv) A is an antisymmetric algebra if and only if B is an antisymmetric algebra.
(v) Every maximal A-antisymmetric set is B-antisymmetric.
(vi) A subset K of X is a maximal A-antisymmetric set if and only if K is a maximal B-antisymmetric set.

Proof. Recall that there is no distinction between partially antisymmetric and antisymmetric sets in the case of a complex function algebra. Hence, in order to prove that a set is B-antisymmetric, it is sufficient to prove that the set is partially B-antisymmetric.

(i) This is obvious since $A \subset B$.

(ii) Note that for x in X and h in B, $\sigma(h)(\tau(x)) = \bar{h}(x)$. Thus h is real (respectively, constant) on K if and only if $\sigma(h)$ is real (respectively, constant) on $\tau(K)$. Since $\sigma: B \to B$ is onto, (ii) follows.

(iii) Let $h \in B$ be real on K. Then $\sigma(h)$ is real on $\tau(K) = K$. Thus

$$f := \frac{h + \sigma(h)}{2} \in A \text{ is real on } K$$

and

$$g := \frac{h - \sigma(h)}{2i} \in A \text{ satisfies } \text{Re}(g_{|K}) \equiv 0.$$

Since K is A-antisymmetric, f and g are constants on K. Hence $h = f + ig$ is a constant on K.

(iv) This follows by letting $K = X$ in (i) and (iii).

(v) Let K be a maximal A-antisymmetric set. Then, by Theorem 2.1.10(b)(ii), $K = \tau(K)$, or K and $\tau(K)$ are disjoint. If $K = \tau(K)$, then K is B-antisymmetric by (iii). If K and $\tau(K)$ are disjoint, by Theorem 2.1.10(b)(ii) there exists h in A such that $h_{|K} \equiv i$. Now, let $f, g \in A$ be such that $f + ig \in B$ is real on K. But $(f + ig)_{|K} = (f + hg)_{|K} = $ constant, since $f + hg \in A$ and K is A-antisymmetric. Thus K is B-antisymmetric.

(vi) Let K be a maximal A-antisymmetric set. Then K is B-antisymmetric by (v). If $K \subset H$ and H is B-antisymmetric, then H is A-antisymmetric by (i). Hence $K = H$. Thus K is a maximal B-antisymmetric set. Now suppose that K is a maximal B-antisymmetric set. Then K is A-antisymmetric by (i). By Theorem 2.1.5 and Remark 2.1.9, K is contained in a maximal A-antisymmetric set T. But then by (v), T is B-antisymmetric and hence $K = T$. Thus K is a maximal A-antisymmetric set. □

In general, we do not know whether the converse of statement (i) in Theorem 2.1.17 is true, that is, whether every A-antisymmetric set is B-antisymmetric. Statements (iii) and (iv) give sufficient conditions for an A-antisymmetric set to be B-antisymmetric.

2.2 ANTISYMMETRIC SETS AND PEAK SETS

One important consequence of Bishop's theorem (Corollary 2.1.6) is that it reduces the study of function algebras to antisymmetric function algebras. To achieve this effectively, the restriction algebras $A_{|K}$, where K is an antisymmetric set, should also be uniformly closed. For this purpose we shall need the concept of a peak set.

Definition 2.2.1 Let X be a compact Hausdorff space and A be a real subalgebra of $C(X)$. A nonempty subset S of X is called an A-peak set (or a peak set for A) if there exists f in A such that $S = \{x \in X : f(x) = 1\}$ and $|f(y)| < 1$ for all y in $X \setminus S$; the function f is said to peak on S. S is called a weak A-peak set (or A-peak set in the weak sense or weak peak set for A) if S is the intersection of some collection of A-peak sets.

As an example, let $X = [0,1]$. Then $f(t) = t$ peaks on $\{1\}$. Thus $\{1\}$ is a peak set for any subalgebra of $C([0,1])$ containing f.

Remark 2.2.2
(i) Note that if f peaks on S, g peaks on T and $S \cap T \neq \varnothing$, then $\alpha f + (1 - \alpha)g \, (0 < \alpha < 1)$ and fg peak on $S \cap T$.

(ii) Let X and A be as in Definition 2.2.1. Then every A-peak set as well as every weak A-peak set is compact. Further, if τ is a topological involution on X and A is a subalgebra of $C(X,\tau)$, then every A-peak set (or weak A-peak set) S is invariant under τ; that is, $\tau(S) = S$.

Theorem 2.2.3 Let X be a compact Hausdorff space and A be a uniformly closed real subalgebra of $C(X)$.
(i) A nonempty intersection of a countable family of A-peak sets is an A-peak set.
(ii) Let S be a weak A-peak set, U an open neighborhood of S, and $\varepsilon > 0$. Then there exists f in A such that $\|f\| = 1, f_{|S} \equiv 1$, and $\|f\|_{X\backslash U} < \varepsilon$.
(iii) Let S be a weak A-peak set. Then for every f in A and $\varepsilon > 0$, there exists h in A such that $h = f$ on S and $\|h\| < \|f\|_S + \varepsilon$.
(iv) If S is a weak A-peak set, then $A_{|S} := \{f_{|S} : f \in A\}$ is uniformly closed in $C(S)$.

Proof.
(i) Let $\{S_n : n = 1,2,\ldots\}$ be a countable family of A-peak sets and $S = \cap\, S_n$ be nonempty. Suppose that f_n peaks on S_n for each n. Then, clearly, $f := \Sigma_n f_n/2^n \in A$, and f peaks on S.
(ii) Let $S = \cap_{\alpha \in \Lambda} S_\alpha$, where each S_α is an A-peak set. Now $\{X\backslash S_\alpha : \alpha \in \Lambda\}$ is an open cover of the compact set $X\backslash U$. Hence $X\backslash U$ is a subset of $(X\backslash S_{\alpha_1}) \cup \cdots \cup (X\backslash S_{\alpha_n})$ for some $S_{\alpha_1},\ldots,S_{\alpha_n}$, so that $\cap_{i=1}^n S_{\alpha_i} \subset U$. By (i), $\cap_{i=1}^n S_{\alpha_i}$ is an A-peak set. Thus there exists f in A such that f peaks on $\cap_{i=1}^n S_{\alpha_i}$. Since $S \subset \cap_{i=1}^n S_{\alpha_i} \subset U$, we have $f_{|S} \equiv 1 = \|f\|$ and $|f(y)| < 1$ for y in $X\backslash U$. Since $X\backslash U$ is compact, $\|f\|_{X\backslash U} < 1$. Now we can consider a sufficiently high power of f, if necessary, to obtain $\|f\|_{X\backslash U} < \varepsilon$.
(iii) We may assume that $\|f\| \leq 1$. Let $U := \{x \in X : |f(x)| < \|f\|_S + \varepsilon\}$. Clearly, U is an open neighborhood of S. Hence by (ii), there is g in A such that $\|g\| = 1, g \equiv 1$ on S, and $\|g\|_{X\backslash U} < \varepsilon$. Let $h = fg \in A$. Clearly, $h = f$ on S. Now, for x in $X\backslash U$, $|h(x)| = |f(x)| |g(x)| < \varepsilon$. Thus $\|h\| < \|f\|_S + \varepsilon$.
(iv) Let $A(S) := \{f \in A : f_{|S} \equiv 0\}$ and $A/A(S)$ be the quotient space. $A/A(S)$ is complete in the quotient norm. For g in A, $g + A(S) \in A/A(S)$ and its quotient norm is given by

$$\||g + A(S)\|| := \inf\{\|g + f\| : f \in A(S)\}$$
$$= \inf\{\|h\| : h \in A, h_{|S} = g_{|S}\}$$
$$= \|g\|_S, \quad \text{by (iii)}.$$

Thus $A_{|S}$ is isometric to $A/A(S)$. Since $A_{|S}$ is complete, it is uniformly closed in $C(S)$. $\qquad\square$

The following theorem gives a relationship between antisymmetric sets and peak sets.

Theorem 2.2.4 Let X be a compact Hausdorff space and A a uniformly closed real subalgebra of $C(X)$ containing 1. Then every maximal partially A-antisymmetric set is a weak peak set for A.

Proof. Let K be a maximal partially A-antisymmetric set and $\mathscr{F}$ the family of all A-peak sets that contain K. Since X is an A-peak set and $K \subset X$, we have $X \in \mathscr{F}$. $\mathscr{F}$ is thus nonempty.

Let $Z := \cap\{S : S \in \mathscr{F}\}$. Clearly, Z is a weak A-peak set and $K \subset Z$. We shall show that Z is a partially A-antisymmetric set. Then the maximality of K implies that $Z = K$ and the conclusion follows.

Suppose that Z is not partially A-antisymmetric. Then there exists f in A such that $f_{|Z}$ is real and nonconstant. Since K is partially A-antisymmetric, $f_{|K}$ is a constant, say a. Since $f_{|Z}$ is nonconstant, there exists z_0 in $Z \backslash K$ such that $f(z_0) = b \neq a$. We define $g_0 := 1 - (f - a)^2/\|f - a\|_Z^2$. Then $g_0 \in A$, $g_0(Z) \subset [0,1]$, $g_{0|K} \equiv 1$, and $g_0(z_0) < 1$. For $n = 1,2,\dots$, let

$$U_n := \{x \in X : |g_0(x)| < 1 + 2^{-n}\}.$$

Then U_n is an open neighborhood of Z. Hence by (ii) of Theorem 2.2.3, there exists g_n in A such that

$$\|g_n\| = 1, \quad g_{n|z} \equiv 1, \quad \|g_n\|_{X \backslash U_n} < (2^n\|g_0\|)^{-1}. \tag{1}$$

Let

$$g := g_0 \sum_{n=1}^{\infty} 2^{-n} g_n. \tag{2}$$

Then

$$|g(x)| \leq |g_0(x)| \quad \text{for all } x \text{ in } X. \tag{3}$$

Now, if $x \in \cap U_n$, then $|g_0(x)| < 1 + 2^{-n}$ for all n, that is, $|g_0(x)| \leq 1$, and hence, by (3),

$$|g(x)| \leq 1 \quad \text{for all } x \text{ in } \bigcap_{n=1}^{\infty} U_n. \tag{4}$$

Consider x in $(X \backslash \cap U_n)$. Let $U_0 = X$. Since $U_{n+1} \subset U_n$ for all n, there exists $m \geq 1$ such that $x \in U_{m-1}$ and $x \notin U_m$. Then $x \notin U_n$ for all $n \geq m$. Now

$$|g_0(x)g_n(x)| \leq 1 + 2^{-m+1} \quad \text{for } n = 1,2,\dots,m-1. \tag{5}$$

For $n \geq m$, since $x \notin U_n$, (1) implies that

$$|g_0(x)g_n(x)| < 2^{-n} < 2^{-m+1}. \tag{6}$$

Combining (5) and (6) we get

$$|g(x)| \leq \sum_{n=1}^{\infty} 2^{-n}|g_n(x)g_0(x)|$$

$$< (1 + 2^{-m+1}) \sum_{n=1}^{m-1} 2^{-n} + 2^{-m+1} \sum_{n=m}^{\infty} 2^{-n}$$

$$= 1;$$

that is,

$$|g(x)| < 1 \qquad \text{for all } x \text{ in } \left(X \setminus \bigcap_{n=1}^{\infty} U_n\right). \tag{7}$$

Inequalities (4) and (7) imply that $\|g\| \leq 1$. Hence the function $(g + 1)/2$ peaks on the set $S := g^{-1}(\{1\})$.

But $g_{0|K} \equiv 1$ and $g_{n|K} \equiv 1$ for all n by (1). Hence, for all x in K,

$$g(x) = g_0(x) \sum_{n=1}^{\infty} 2^{-n}g_n(x) = 1.$$

Thus $K \subset S$. This means that S is an A-peak set containing K. Hence $S \in \mathcal{F}$. Then $Z \subset S = g^{-1}(\{1\})$. But $z_0 \in Z$ and from (3) and the choice of g_0,

$$|g(z_0)| \leq |g_0(z_0)| < 1, \qquad \text{which is a contradiction.} \qquad \square$$

Corollary 2.2.5
(i) Let B be a complex function algebra on a compact Hausdorff space X and K a maximal B-antisymmetric set. Then $B_{|K}$ is uniformly closed in $C(K)$.
(ii) Let τ be a topological involution on X, A a real function algebra on (X,τ), and H a maximal A-antisymmetric set. Then $A_{|H}$ is uniformly closed in $C(H)$.

Proof.
(i) Since B contains complex constants, K is also a maximal partially B-antisymmetric set. (See the comments following Definition 2.1.8.) Hence by Theorem 2.2.4, K is a weak peak set for B. Now (iv) of Theorem 2.2.3 implies that $B_{|K}$ is uniformly closed in $C(K)$.
(ii) By Theorem 2.1.10(b)(ii), $H \cup \tau(H)$ is a maximal partially A-

antisymmetric set. Hence $H \cup \tau(H)$ is a weak peak set for A by Theorem 2.2.4. Again, (iv) of Theorem 2.2.3 implies that $A|_{H \cup \tau(H)}$ is uniformly closed in $C(H \cup \tau(H))$. Further, for each g in A, $\|g\|_H = \|g\|_{H \cup \tau(H)}$ by Remark 2.1.11. Thus $A|_H$ is isometric to $A_{|H \cup \tau(H)}$, and hence closed in $C(H)$. $\qquad\square$

Definition 2.2.6 Let X be a compact Hausdorff space. A real subspace A of $C(X)$ is said to be *extremely regular at a subset S of X* if for every open neighborhood U of S and for each $\varepsilon > 0$, there is f in A such that:
(i) $f_{|S} \equiv 1 = \|f\|$.
(ii) $|f(y)| < \varepsilon$ for all y in $X \backslash U$.

Lemma 2.2.7 [Compare Lemma 1 of Jarosz (1984b).] Let X be a compact Hausdorff space and A be a real subspace of $C(X)$. Suppose that A is extremely regular at a subset S of X. Then for every $\varepsilon > 0$ and for every open neighborhood U of S, there is a function f in A such that
(i) $f_{|S} \equiv 1 = \|f\|$.
(ii) $|f(y)| < \varepsilon$ for all y in $X \backslash U$.
(iii) $|\operatorname{Im} f(x)| < \varepsilon$ for all x in X.

Proof. Let $\varepsilon > 0$, U be an open neighborhood of S and n be a natural number such that $1/n < \varepsilon/2$. Let $U_1 := U$ and f_1 be a function in A such that $f_{1|S} \equiv 1 = \|f_1\|$ and $|f_1(x)| < \varepsilon/2$ for all x in $X \backslash U_1$.
Having defined U_K and f_k, let

$$U_{k+1} := \left\{ x \in U_k : |f_k(x) - 1| < \frac{\varepsilon}{2} \right\}$$

and f_{k+1} be an element of A with $\|f_{k+1}\| = 1 \equiv f_{k+1|S}$ and $|f_{k+1}(x)| < \varepsilon/2$ for all x in $X \backslash U_{k+1}$.
We define $f := (1/n) \Sigma_{k=1}^n f_k$. Obviously, f satisfies (i) and (ii) as each f_k does. If $x \in X \backslash U_1$, then for all k, $|f_k(x)| < \varepsilon/2$, and consequently, $|f(x)| < \varepsilon/2$, so that (iii) is satisfied. If $x \in U_1$, denote by k_0 the greatest positive integer not greater than n such that $x \in U_{k_0}$. Then

$$\left| f(x) - \frac{k_0 - 1}{n} \right| = \left| \frac{1}{n} \left[\sum_{j=1}^{k_0-1} (f_j(x) - 1) + f_{k_0}(x) + \sum_{j=k_0+1}^{n} f_j(x) \right] \right|$$

$$\leq \frac{1}{n} \left[(k_0 - 1) \frac{\varepsilon}{2} + 1 + (n - k_0) \frac{\varepsilon}{2} \right]$$

$$= \frac{n - 1}{n} \frac{\varepsilon}{2} + \frac{1}{n}$$

$$< \varepsilon.$$

Hence

$$\left| \operatorname{Im} f(x) \right| = \left| \operatorname{Im} \left(f(x) - \frac{k_0 - 1}{n} \right) \right| \le \left| f(x) - \frac{k_0 - 1}{n} \right| < \varepsilon. \quad \square$$

We now give some characterizations of weak peak sets.

Theorem 2.2.8 Let X be a compact Hausdorff space, A a uniformly closed real subalgebra of $C(X)$ containing 1, and S a subset of X. Then the following statements are equivalent.
(i) S is a weak peak set for A.
(ii) A is extremely regular at S.
(iii) For every $\varepsilon > 0$ and for every open neighborhood U of S, there exists a function f in A such that $f|_S \equiv 1 = \|f\|$, $|f(y)| < \varepsilon$ for all y in $X \backslash U$, and $|\operatorname{Im} f(x)| < \varepsilon$ for all x in X.
(iv) For every open neighborhood U of S, there exists an A-peak set T such that $S \subset T \subset U$.
(v) There exist constants M and c with $0 < c < 1 \le M$ such that for every open neighborhood U of S, there exists a function f in A with the following properties: $\|f\| \le M$, $f|_S \equiv 1$, and $|f(y)| < c$ for all y in $X \backslash U$.

Proof. "(i) implies (ii)" follows from Theorem 2.2.3(ii) and "(ii) implies (iii)" follows from Lemma 2.2.7.

(iii) implies (iv): Let $0 < \varepsilon < 1$ and U be an open neighborhood of S. By (iii) we can find f in A, such that $f|_S \equiv 1 = \|f\|$ and $|f(y)| < \varepsilon$ for all y in $X \backslash U$. Let $T := \{x \in X : f(x) = 1\}$. Then $S \subset T \subset U$ and the function $g := (1 + f)/2$ belongs to A and peaks on T.

(iv) implies (i): Let Λ be the family of all open neighborhoods of S. For each U in Λ, we obtain an A-peak set T_U such that $S \subset T_U \subset U$. Now

$$S \subset \cap \{T_U : U \in \Lambda\} \subset \cap \{U : U \in \Lambda\} = S.$$

Thus $S = \cap \{T_U : U \in \Lambda\}$ is a weak A-peak set.

The argument above proves the equivalence of (i), (ii), (iii), and (iv). Further, (ii) obviously implies (v). Hence to complete the proof, we establish "(v) implies (ii)." If $M = 1$, then (ii) follows by replacing the function f in (v) by a large enough power of f. Next, let $M > 1$.

Let β be a real number such that $1 < \beta < (M - c)(M - 1)^{-1}$. Choose $\delta > 0$ in such a way that $\delta < M - c - \beta(M - 1)$. Let U be an open neighborhood of S and $U_1 = U$. Then there exists f_1 in A with $\|f_1\| \le M$, $f_{1|S} \equiv 1$, and $|f_1(y)| < c$ for all y in $X \backslash U_1$.

Having defined open neighborhoods $U_1, \ldots, U_n$ of S, and $f_1, \ldots, f_n$, define U_{n+1} by

$$U_{n+1} := \{x \in U_n : |f_j(x)| < 1 + \delta\beta^{-(n+1)}, 1 \le j \le n\}.$$

Then U_{n+1} is an open neighborhood of S. Hence there exists f_{n+1} in A such that $\|f_{n+1}\| \le M$, $f_{n+1|_S} \equiv 1$, and $|f_{n+1}(y)| < c$ for all y in $X\backslash U_{n+1}$. Let $f = (\beta - 1)\Sigma_{n=1}^{\infty}\beta^{-n}f_n$. Then $f_{|_S} \equiv 1$.

If $y \notin U = U_1$, then $y \notin U_n$, so that $|f_n(y)| < c$ for all n. Thus $|f(y)| < c(\beta - 1)\Sigma_{n=1}^{\infty}\beta^{-n} = c < 1$.

We now prove that $\|f\| \le 1$. Since we already know that $|f(y)| < 1$ for $y \notin U$, it is sufficient to show that $|f(y)| \le 1$ for all y in U. Suppose that $y \in U = U_1$.

Case (i): Let $y \in \cap_{n=1}^{\infty} U_n$. In this case, $|f_j(y)| < 1 + \delta\beta^{-n}$ for all $n > j$, which implies that $|f_j(y)| \le 1$ for all j. Hence $|f(y)| \le 1$.

Case (ii): There exists n such that $y \in U_n \backslash U_{n+1}$. Then $y \notin U_j$ for $j > n$. In this case,

$$|f(y)| \le (\beta - 1)\left[\sum_{j=1}^{n-1} |f_j(y)|\beta^{-j} + |f_n(y)|\beta^{-n} + \sum_{j=n+1}^{\infty} \beta^{-j}|f_j(y)|\right]$$

$$\le (\beta - 1)\left[\sum_{j=1}^{n-1} (1 + \delta\beta^{-n})\beta^{-j} + M\beta^{-n} + c\sum_{j=n+1}^{\infty} \beta^{-j}\right]$$

$$= 1 - \beta^{-n}(\beta - \delta - M(\beta - 1) - c + \delta\beta^{1-n})$$

$$< 1 \quad \text{as} \quad \beta - M(\beta - 1) - c - \delta > 0.$$

This proves that $\|f\| \le 1$.

Let $\varepsilon > 0$ be given. We may consider a sufficiently high power of f (in place of f) to obtain

$$|f(y)| < \varepsilon \quad \text{for all } y \text{ in } X\backslash U.$$

This proves (ii). $\qquad\qquad\qquad\qquad\qquad\qquad\qquad\qquad\qquad\qquad$ $\square$

Jarosz (1984a) proved that if A is a complex function algebra on X, then each of the statements in Theorem 2.2.8 is also equivalent to the following: There exist $M \ge 1$, $c_1 \ge 0$, $c_2 \ge 0$ with $c_1 + c_2 < 1$ such that for every open neighborhood U of S, there exists f in A with $\|f\| \le M$, $|f(s) - 1| \le c_1$ for all s in S, and $|f(y)| \le c_2$ for all y in $X\backslash U$. Arundhathi (1988) used this theorem of Jarosz along with the complexification technique to prove a similar result in the case of a real function algebra.

Remark 2.2.9 Let X and A be as in Theorem 2.2.8. Every A-peak set is G_δ: If f peaks on S, then $S = \cap_n G_n$, where $G_n := \{x \in X : |f(x)| > 1 - 1/n\}$, $n = 1, 2, \ldots$. Further, Theorem 2.2.8 shows that if a weak peak set is G_δ, then it is a peak set: Let S be a weak A-peak set. If S is G_δ, then $S = \cap_n G_n$, for some countable family of open sets G_n, $n = 1,2,\ldots$.

Now, by Theorem 2.2.8(iv), for each n there exists a peak set T_n such that $S \subset T_n \subset G_n$. Hence $S = \cap T_n$ is a peak set by Theorem 2.2.3(i). In particular, if X is metrizable, then every closed subset of X is a G_δ-set, so that every weak A-peak set is a peak set for A.

Remark 2.2.10 Let X be a compact Hausdorff space and A a uniformly closed real subalgebra of $C(X)$ containing 1. Let S and T be A-peak sets. Then $S \cup T$ is also an A-peak set. The following proof of this fact due to Bear is given in Browder (1969). Let f peak on S and g peak on T. Since the function $(1 - z)^{1/2}$ (principal value) can be approximated uniformly on the closed unit disk by polynomials with real coefficients, the spectral mapping theorem (Theorem 1.1.26) shows that the functions $h := (1 - f)^{1/2}$ and $k := (1 - g)^{1/2}$ belong to A. hk vanishes precisely on $S \cup T$. Also, since $|\arg h| < \pi/4$ and $|\arg k| < \pi/4$, Re $hk > 0$ outside $S \cup T$. Thus $\exp(-hk) \in A$ and peaks on $S \cup T$.

Now let S and T be weak A-peak sets and U be an open neighborhood of $S \cup T$. By Theorem 2.2.8(iv), there exists A-peak sets P and Q such that $S \subset P \subset U$ and $T \subset Q \subset U$. By the argument above, $P \cup Q$ is an A-peak set and $S \cup T \subset P \cup Q \subset U$. Hence again by Theorem 2.2.8, $S \cup T$ is a weak A-peak set.

We now investigate the relationship between the peak sets for a real function algebra and those of its complexification.

Theorem 2.2.11 Let X be a compact Hausdorff space, τ a topological involution on X, A a uniformly closed real subalgebra of $C(X)$ containing 1, and $B := \{f + ig : f,g \in A\}$.

(i) Every A-peak set (respectively, weak A-peak set) is a B-peak set (respectively, weak B-peak set).

(ii) If S is a B-peak set (respectively, weak B-peak set), then $\tau(S)$ is a B-peak set (respectively, weak B-peak set). In case $\tau(S) = S$, S is, in fact, an A-peak set (respectively, weak A-peak set).

(iii) If S is a B-peak set (respectively, weak B-peak set), then $S \cup \tau(S)$ is an A-peak set (respectively, weak A-peak set).

(iv) Let $x \in X$. If $\{x\}$ is a B-peak set (respectively, weak B-peak set), then $\{x,\tau(x)\}$ is an A-peak set (respectively, weak A-peak set). Let $\{x,\tau(x)\}$ be an A-peak set (respectively, a weak A-peak set). If $\tau(x) = x$, then $\{x\}$ is a B-peak set (respectively, a weak B-peak set). If $\tau(x) \neq x$, and A separates x and $\tau(x)$, then again $\{x\}$ is a B-peak set (respectively, a weak B-peak set).

Proof.

(i) If S is an A-peak set, there is f in A that peaks on S. Clearly, $f \in B$. Hence S is a B-peak set. If S is a weak A-peak set, then $S = \cap \{S_\alpha : \alpha \in \Lambda\}$, where each S_α is an A-peak set. But then each S_α is a B-peak set. Hence S is a weak B-peak set.

(ii) Suppose that $h \in B$ peaks on S. Since $\sigma(h) \in B$ and $\sigma(h)(\tau(x)) = \bar{h}(x)$ for all x in X, it is easy to prove that $\sigma(h)$ peaks on $\tau(S)$, thus making it a B-peak set.

If $S = \tau(S)$, h and $\sigma(h)$ both peak on S. Hence $(h + \sigma(h))/2$ as well as $h\sigma(h)$ peak on S and they belong to A. Hence S is an A-peak set.

Now if S is a weak B-peak set, then $S = \cap \{S_\alpha : \alpha \in \Lambda\}$, where each S_α is a B-peak set. Then $\tau(S) = \cap \{\tau(S_\alpha) : \alpha \in \Lambda\}$, and each $\tau(S_\alpha)$ is a B-peak set, making $\tau(S)$ a weak B-peak set.

If $S = \tau(S)$, then $S = \cap \{S_\alpha \cap \tau(S_\alpha) : \alpha \in \Lambda\}$. Each $\tau(S_\alpha)$ is a B-peak set. Further, $S_\alpha \cap \tau(S_\alpha)$ is nonempty as it contains S. Hence $S_\alpha \cap \tau(S_\alpha)$ is a B-peak set. Since it is invariant under τ, it is also an A-peak set. As this is true for each α, S is a weak A-peak set.

(iii) Let S be a B-peak set. Then $\tau(S)$ is a B-peak set by (ii), and hence $S \cup \tau(S)$ is a B-peak set by Remark 2.2.10. Since $\tau(S \cup \tau(S)) = S \cup \tau(S)$, $S \cup \tau(S)$ is an A-peak set by (ii). Similar argument holds if S is a weak B-peak set.

(iv) The first part follows from (iii) with $S = \{x\}$.

Now let $\{x, \tau(x)\}$ be a weak A-peak set. Then it is a weak B-peak set by (i). If $x = \tau(x)$, then clearly x is a weak B-peak set. Now let $\tau(x) \neq x$, and assume that A separates x and $\tau(x)$. Then so does B. Since B is a complex algebra, there exists g in B such that $g(x) = 1$ and $g(\tau(x)) = 0$. In view of Theorem 2.2.3(iii), we may assume that $\|g\| \leq 2$. Let V be an open neighborhood of $\{x\}$ and let $\varepsilon > 0$. Let $W := \{y \in X : |g(y)| < \varepsilon\}$. Then $V \cup W$ is a neighborhood of $\{x, \tau(x)\}$. Hence by Theorem 2.2.8(iii) there exists h in B such that $h(x) = 1 = h(\tau(x))$, $\|h\| = 1$, and $|h(y)| < \varepsilon/2$ for all y in $X \setminus (V \cup W)$. Now consider $f := gh$ in B. Then $f(x) = 1$ and $\|f\| \leq 2$. Let $y \in X \setminus V$. If $y \in X \setminus (V \cup W)$, then $|f(y)| = |g(y)| |h(y)| < \varepsilon$ as $|g(y)| \leq 2$, and $|h(y)| < \varepsilon/2$. If $y \in W$, then $|f(y)| = |g(y)| |h(y)| < \varepsilon$ as $|g(y)| < \varepsilon$ and $|h(y)| \leq 1$. Thus for all y in $X \setminus V$, $|f(y)| < \varepsilon$. Now from Theorem 2.2.8(v), it follows that $\{x\}$ is a weak B-peak set.

Next let $\{x, \tau(x)\}$ be a peak set for A. Then it is a B-peak set by (i). If $\tau(x) = x$, then clearly $\{x\}$ is a B-peak set. Let, then, $\tau(x) \neq x$. Now $\{x\}$ is a weak B-peak set by the argument above. Hence $\{x\} = \cap S_\alpha$ for some family of B-peak sets S_α. At least one S_α does not contain $\tau(x)$. Hence for this α, $\{x\} = S_\alpha \cap \{x, \tau(x)\}$. Thus $\{x\}$ is a B-peak set by Theorem 2.2.3(i). $\qquad\square$

Example 2.2.12 Let B be the (complex) disk algebra of Example 1.3.10. For each s with $|s| = 1$, the function $(1 + \bar{s}z)/2$ peaks on $\{s\}$. Thus $\{s\}$ is a B-peak set. Since B can be viewed as a complexification of the real disk algebra A of Example 1.3.10, we can apply Theorem 2.2.11 to conclude that $\{s, \bar{s}\}$ is an A-peak set for each s with $|s| = 1$; of course, if $s = \bar{s}$, that is, $s = \pm 1$, then $\{s\}$ is an A-peak set.

Some properties of the peak sets and antisymmetric sets discussed in this section were proved in Kulkarni and Srinivasan (1987, 1988a), Arundhathi (1990), and Hwang (1990). The last two references contain some additional material about peak sets.

2.3 HOFFMAN–WERMER THEOREM AND ITS ANALOG

Let X be a compact Hausdorff space and A a subset of $C(X)$. We use the following notation.

$$\text{Re } A := \{\text{Re } f : f \in A\}.$$

Clearly, if A is a subspace of $C(X)$, then $\text{Re } A$ is a real subspace of $C_{\mathbb{R}}(X)$. What additional properties does $\text{Re } A$ have if A is a (real or complex) function algebra? We now begin the investigation of this question. In the case of complex algebras such an investigation started with the classical theorem of Hoffman and Wermer, which says that $\text{Re } A$ cannot be closed under uniform convergence unless $A = C(X)$. In other words, if $\text{Re } A$ is uniformly closed, then $A = C(X)$ (Hoffman and Wermer, 1962). The classical proof of this theorem made a crucial use of Bishop's theorem (Corollary 2.1.6) as follows. It was shown that if $\text{Re } A$ is uniformly closed and A is antisymmetric, then X is a singleton set. This was followed by showing that if $\text{Re } A$ is uniformly closed (for an arbitrary complex function algebra), then $\text{Re}(A_{|K})$ is uniformly closed in $C_{\mathbb{R}}(K)$ for every maximal A-antisymmetric set K. Since $A_{|K}$ is a complex function algebra (Corollary 2.2.5), these two facts imply that every maximal A-antisymmetric set is a singleton, and the conclusion follows from Bishop's theorem.

In this section we prove the following analog of the Hoffman–Wermer theorem. Let τ be a topological involution on X and A be a real function algebra on (X, τ). If $\text{Re } A$ is uniformly closed, then $A = C(X, \tau)$. We have to make a few modifications in the technique of Hoffman and Wermer, because $A_{|K}$ may not be a real function algebra for every maximal A-antisymmetric set K, as $\tau_{|K}$ may not be an involution on K. The classical

Hoffman–Wermer theorem is also proved in the course of proving its real analog.

Lemma 2.3.1 Let X be a compact Hausdorff space and A, a uniformly closed real subalgebra of $C(X)$ that contains real constants and separates the points of X. If A is an antisymmetric algebra and Re A is uniformly closed, then X is a singleton set.

Proof. We define a map T: Re $A \to A$ as follows: Let $u \in$ Re A and $x_0 \in X$.

Case (i): If A contains the constant function i, then there exists a unique f in A such that Re $f = u$ and Im $f(x_0) = 0$. (The uniqueness follows by the antisymmetry of A.) Define $T(u) = f$.

Case (ii): Suppose that A does not contain the constant function i. If for f_1, f_2 in A, Re $f_1 = u =$ Re f_2, then Re$(f_1 - f_2) = 0$. Hence by the antisymmetry of A, $f_1 - f_2$ is a constant, and this constant must be 0, as i does not belong to A.

Hence there exists a unique f in A such that Re $f = u$. Define $T(u) = f$. Then T is a real linear map. Now suppose that $u_n \in$ Re A, $T(u_n) = f_n$ for $n = 1, 2, \ldots$ and $u_n \to u$, $T(u_n) \to f$, as $n \to \infty$. Then, clearly, $\|\text{Re } f_n - \text{Re } f\| \le \|f_n - f\| \to 0$; that is, $u_n \to$ Re f, as $n \to \infty$. Hence $u =$ Re f. Further, if T is as in case (i), then Im $f_n(x_0) = 0$ for each n. Hence Im $f(x_0) = 0$. Thus $T(u) = f$ in both cases. This shows that T has a closed graph. Hence T is a bounded linear map.

Suppose that X contains more than one point. Since A separates the points of X, there exists a nonconstant function $g = u + iv$ in A [such that $g(x_0) = 0$, in case (i)]. Since g is nonconstant and A is antisymmetric, $u \ne 0$ and $v \ne 0$. Hence the rectangle

$$R := \{s + it \in \mathbb{C} : |s| \le \|u\|, \ |t| \le \|v\|\}$$

has a nonempty interior. Now there exists x_1 in X such that $|v(x_1)| = \|v\|$. Let $z_1 := g(x_1)$. Then z_1 and $\bar{z}_1$ are boundary points of R. Let $M > 0$ and

$$R_M := \{s + it \in \mathbb{C} : |s| \le \|u\|, \ |t| \le M\}.$$

Note that R and R_M are symmetric about the real axis (see Definition 1.1.34). There exists a conformal map ϕ of the interior of R onto the interior of R_M such that the map extends to a homeomorphism of boundaries of R and R_M in such a way that $\phi(z_1) = iM$ and $\phi(\bar{z}_1) = -iM$ [see Rudin (1966, p. 282)]. Then ϕ is a uniform limit of a sequence of polynomials $\{p_n(z)\}$ on R. Define $\psi: R \to R_M$ by $\psi(z) := (\phi(z) + \overline{\phi(\bar{z})})/2$, and let $q_n(z) := (p_n(z) + \overline{p_n(\bar{z})})/2$. Then $\{q_n\}$ converges to ψ uniformly on R. Further, since each q_n is a polynomial with real coefficients, $q_n(g) \in A$ for each n, and $\{q_n(g)\}$ converges to $\psi(g)$ uniformly on X. Hence $h := \psi(g) \in A$ (see Theorem 1.1.26). Also, in case (i), $h(x_0) = \psi(g(x_0)) = \psi(0) = (\phi(0) + \overline{\phi(0)})/2$, so that Im $h(x_0) = 0$. Thus $T(\text{Re } h) = h$. Further,

$$h(x_1) = \psi(g(x_1)) = \psi(z_1) = \frac{\phi(z_1) + \overline{\phi(\bar{z}_1)}}{2}$$

$$= \frac{iM + \overline{(-iM)}}{2} = iM.$$

Thus we see that $\|\text{Re } h\| \leq \|u\|$, while $\|h\| \geq M$. Since M is arbitrary, this contradicts the fact that T is a bounded linear map. Hence X is a singleton set. □

Lemma 2.3.2 Let X be a compact Hausdorff space. Let A be a uniformly closed real subalgebra of $C(X)$ containing 1, and let S be a weak peak set for A. If Re A is uniformly closed, then $\text{Re}(A_{|S}) := \{\text{Re } f_{|S} : f \in A\}$ is uniformly closed in $C_{\mathbb{R}}(S)$.

Proof. [Compare this with the proof of (iv) of Theorem 2.2.3.] First we prove the following: If $f \in A$ and $\varepsilon > 0$, there exists $h \in A$ such that $\|\text{Re } h\| \leq \|\text{Re } f\|_S + 2\varepsilon$ and $h = f$ on S. We may assume that $\|f\| \leq 1$. Let

$$U := \{x \in X : |\text{Re } f(x)| < \|\text{Re } f\|_S + \varepsilon\}.$$

Then U is an open neighborhood of S. Hence, by Theorem 2.2.8, there exists g in A such that $\|g\| = 1 \equiv g_{|S}$, $\|g\|_{X \setminus U} < \varepsilon$, and $|\text{Im } g(x)| < \varepsilon$ for all x in X. Let $h := fg \in A$. Since $g = 1$ on S, we have $h = f$ on S.
   Further, Re $h = \text{Re } f \text{ Re } g - \text{Im } f \text{ Im } g$. Now on U,

$$|\text{Re } h| \leq |\text{Re } f| |\text{Re } g| + |\text{Im } f| |\text{Im } g|$$
$$< (\|\text{Re } f\|_S + \varepsilon) + \varepsilon,$$

while, on $X \setminus U$, $|\text{Re } h| < \varepsilon + \varepsilon$. Thus $\|\text{Re } h\| \leq \|\text{Re } f\|_S + 2\varepsilon$ and $h = f$ on S. This proves the claim. Now, to prove that $\text{Re}(A_{|S})$ is uniformly closed, consider

$$R_S := \{u \in \text{Re } A : u_{|S} \equiv 0\}.$$

Since Re A is assumed to be uniformly closed in $C_{\mathbb{R}}(X)$, it is a real Banach space under the supremum norm. R_S is a closed subspace of Re A. Hence the quotient space Re A/R_S is complete in the quotient norm. Also, the quotient norm of an element Re $f + R_S$ in Re A/R_S is given by

$$\|\|\text{Re } f + R_S\|\| := \inf\{\|\text{Re } h\| : h \in A \text{ and Re } h = \text{Re } f \text{ on } S\}.$$

Since for every $\varepsilon > 0$, there is h in A with $\|\text{Re } h\|_S \leq \|\text{Re } h\| \leq \|\text{Re } f\|_S$

$+ 2\varepsilon$ and Re h = Re f on S, we see that $\|\|\mathrm{Re}\, f + R_S\|\| = \|\mathrm{Re}\, f\|_S$. Thus $\mathrm{Re}(A_{|S})$ is linearly isometric to Re A/R_S and is complete. Hence $\mathrm{Re}(A_{|S})$ is uniformly closed.   □

Theorem 2.3.3 (Hoffman–Wermer theorem) Let X be a compact Hausdorff space and A a subset of $C(X)$ such that Re A is uniformly closed.
(i) If A is a complex function algebra on X, then $A = C(X)$.
(ii) If τ is a topological involution on X and A is a real function algebra on (X,τ), then $A = C(X,\tau)$.

Proof. In both cases, A is a uniformly closed real subalgebra of $C(X)$ that contains 1 and separates the points of X. We shall show that every maximal A-antisymmetric set is a singleton. Then the conclusion will follow by Bishop's theorem (Corollary 2.1.6) in case (i), and by its analog (Corollary 2.1.13) in case (ii). So let K be a maximal A-antisymmetric set. Then $A_{|K}$ is a uniformly closed (Corollary 2.2.5) real subalgebra of $C(K)$ that contains 1 and separates the points of K. Also, $A_{|K}$ is an antisymmetric algebra. Now if we show that $\mathrm{Re}(A_{|K})$ is uniformly closed in $C_\mathbb{R}(K)$, we can apply Lemma 2.3.1 to the algebra $A_{|K}$ to conclude that K is a singleton. Thus, to complete the proof, it remains only to show that $\mathrm{Re}(A_{|K})$ is uniformly closed in both the cases.

  (i) Since A is a complex algebra, K is also a maximal partially A-antisymmetric set. Hence K is a weak peak set for A by Theorem 2.2.4. Hence $\mathrm{Re}(A_{|K})$ is uniformly closed by Lemma 2.3.2.

  (ii) $K \cup \tau(K)$ is a maximal partially A-antisymmetric set by Theorem 2.1.10(b)(ii), and is hence a weak peak set for A by Theorem 2.2.4. Hence $\mathrm{Re}(A_{|K\cup\tau(K)})$ is uniformly closed by Lemma 2.3.2. For each f in A and x in X, $f(\tau(x)) = \bar{f}(x)$, so that Re $f(\tau(x))$ = Re $f(x)$. Therefore, $\|\mathrm{Re}\, f\|_K = \|\mathrm{Re}\, f\|_{\tau(K)} = \|\mathrm{Re}\, f\|_{K\cup\tau(K)}$. Thus $\mathrm{Re}(A_{|K})$ is linearly isometric to $\mathrm{Re}(A_{|K\cup\tau(K)})$ and is hence uniformly closed in $C_\mathbb{R}(K)$.   □

Corollary 2.3.4 Let X be a compact Hausdorff space.
(i) If A is a complex function algebra on X such that Re $A = C_\mathbb{R}(X)$, then $A = C(X)$.
(ii) If τ is a topological involution on X, A a real function algebra on (X,τ) such that Re $A = C_\mathbb{R}(X,\tau)$, then $A = C(X,\tau)$.

Proof. Obvious.   □

  Bishop's theorem and its analog can be used to prove the following theorem.

Theorem   2.3.5   Let X be a compact metric space.
(i) (Gorin, 1965) Let B be a uniformly closed complex subalgebra of $C(X)$ that contains 1. If every strictly positive function in $C_\mathbb{R}(X)$ is the modulus (absolute value) of an invertible element of B, then $B = C(X)$.
(ii) (Kulkarni and Srinivasan, 1988b) Let τ be a topological involution on X and A a real function algebra on (X,τ). If every strictly positive function in $C_\mathbb{R}(X,\tau)$ is the modulus of an invertible element of A, then $A = C(X,\tau)$.

The details of the proof can be found in the references cited above. We omit the proofs as the technique is similar to the proof of Theorem 2.3.3. Gorin (1965) also gives an example to show that the hypothesis of metrizability of X cannot be dropped. This example and a simplification of Gorin's proof can be found in Stout (1971).

Let X be a compact Hausdorff space and Y a closed subset of X. Sidney and Stout (1968) have proved that if A is a complex function algebra on X and if Re $A_{|Y}$ is uniformly closed in $C_\mathbb{R}(Y)$, then $A_{|Y} = C(Y)$. A proof of this can be found in Burckel (1972). We do not know whether an analog of this theorem for real function algebras is true. We may formulate an analog as follows.

Let τ be a topological involution on a compact Hausdorff space X, Y a closed subset of X such that $\tau(Y) = Y$ and A a real function algebra on (X,τ). If Re $A_{|Y}$ is uniformly closed, then $A_{|Y} = C(Y,\tau_{|Y})$.

The following generalization of the Hoffman–Wermer theorem [due to Bernard (1968)] is used in proving the above-mentioned result of Sidney and Stout. Let X be a compact Hausdorff space and A a complex Banach function algebra on X; that is, A is contained in $C(X)$, separates the points of X, contains the constants, and is complete in some norm. If Re A is uniformly closed, then $A = C(X)$. It would be of interest to know whether a real analog of this result is true.

Cirka proved in 1966 that if X is a locally connected compact Hausdorff space and A is a complex function algebra on X such that the set $\{g^2 : g \in A\}$ is dense in A, then $A = C(X)$ (Burckel, 1972). It will be interesting to obtain an analog of this result for real function algebras.

Corollary   2.3.6   Let X be a compact Hausdorff space.
(i) (Cerych, 1986) If A is a complex function algebra on X and A has a finite codimension in $C(X)$, then $A = C(X)$.
(ii) (Grzesiak, 1989) If τ is a topological involution on X, A is a real function algebra on (X,τ), and A has a finite codimension in $C(X,\tau)$, then $A = C(X,\tau)$.

Proof.   If $A \neq C(X)$ (respectively, $C(X,\tau)$), then by Remark 2.1.7 (respectively, Corollary 2.1.14), A is not conjugate-closed; that is, there exists f_1 in A such that $\bar{f}_1 \notin A$. Let $A_0 = A$, $A_1 = \text{span}(A \cup \{\bar{f}_1\})$ and $\bar{A} = \{\bar{f} : f \in A\}$. If $\bar{A}$ is not contained in A_1, we can find f_2 in A such that $\bar{f}_2 \notin A_1$. Let $A_2 = \text{span}(A_1 \cup \{\bar{f}_2\})$. Having constructed $A_1, \ldots, A_k$, if $\bar{A}$ is not contained in A_k, we can choose $f_{k+1} \in A$ such that $\bar{f}_{k+1} \notin A_k$ and let $A_{k+1} = \text{span}(A_k \cup \{\bar{f}_{k+1}\})$. Since A has a finite codimension in $C(X)$ [respectively, $C(X,\tau)$], $\bar{A} \subset A_j$ for some natural number j. Then

$$A + \bar{A} = A_j = \text{span}(A \cup \{\bar{f}_1, \bar{f}_2, \ldots, \bar{f}_j\}).$$

Since A is uniformly closed and A has a finite codimension in A_j, A_j is uniformly closed. We shall now show that Re A is uniformly closed. Then the conclusion will follow from the Hoffman–Wermer theorem (Theorem 2.3.3).

Let $u_n \in \text{Re } A$ for $n = 1,2,\ldots$, and $u_n \to u$ as $n \to \infty$. There exists $g_n \in A$ such that $u_n = \text{Re } g_n = (g_n + \bar{g}_n)/2 \in A + \bar{A} = A_j$ for $n = 1,2,\ldots$. Since A_j is closed, $u \in A_j = A + \bar{A}$. Thus $u = f + \bar{g}$ for some f,g in A. But then

$$u = \text{Re } u = \text{Re } f + \text{Re } \bar{g} = \text{Re}(f + g) \in \text{Re } A.$$

Hence Re A is uniformly closed. $\qquad\qquad\qquad\qquad\qquad\qquad\square$

Grzesiak (1989) used the complexification technique along with part (i) of Corollary 2.3.6 to prove part (ii).

2.4 ANALOG OF WERMER'S THEOREM

Wermer proved in 1963 that if A is a complex function algebra on a compact Hausdorff space X and if Re A is a ring, then $A = C(X)$. The original proof of Wermer (1963) was based on Bishop's theorem. He first showed that if Re A is a ring and A is an antisymmetric algebra, then X is a singleton set. Since $A_{|K}$ is an antisymmetric complex function algebra whenever K is a maximal A-antisymmetric set and since $\text{Re}(A_{|K})$ is obviously a ring whenever Re A is a ring, we obtain that every maximal A-antisymmetric set is a singleton. Hence $A = C(X)$ by Bishop's theorem (Corollary 2.1.6). Subsequently, a different proof of Wermer's theorem was given by Bernard. Both proofs can be found in Burckel (1972). In this section we prove the following analog of Wermer's theorem for a real function algebra. Let X be a compact Hausdorff space, τ be a topological involution on X, and A be a real function algebra on (X,τ). If Re A is a ring, then $A = C(X,\tau)$.

Our proof of this analog is similar to Bernard's proof. We do not know whether the original proof of Wermer can be modified to yield this analog.

Definition 2.4.1 Let E and F be normed linear spaces over $\mathbb{R}$ or $\mathbb{C}$. We say that E *is continuously injected in* F if E is a subset of F and the inclusion map is continuous, that is, there is $M > 0$ such that $\|x\|_F \leq M\|x\|_E$ for all x in E, where $\|\cdot\|_E$ and $\|\cdot\|_F$ denote norms on E and F, respectively.

Definition 2.4.2 Let E be a real or complex normed linear space. Let $\tilde{E} := l^\infty(\mathbb{N}, E)$ be the set of all bounded sequences in E. For $f = \{f_n\}$ in $\tilde{E}$, we define $\|f\|_\infty := \sup\{\|f_n\| : n \in \mathbb{N}\}$. Then $(\tilde{E}, \|\cdot\|_\infty)$ is a normed linear space.

The following lemma is crucial in the main proof.

Lemma 2.4.3 (Bernard's lemma) (Burckel, 1972, p. 44) Suppose that E and F are (real or complex) normed linear spaces, and E is continuously injected in F. Then $\tilde{E}$ is continuously injected in $\tilde{F}$. If, in addition, E is complete and $\tilde{E}$ is dense in $\tilde{F}$, then $E = F$.

Proof. Since E is continuously injected in F, there is a constant $M > 0$ such that $\|a\|_F \leq M\|a\|_E$ for every a in E. Hence for f in $\tilde{E}$, $\|f\|_{\tilde{F}} \leq M\|f\|_{\tilde{E}}$. Hence $\tilde{E}$ is continuously injected in $\tilde{F}$.

Now suppose that E is complete and $\tilde{E}$ is dense in $\tilde{F}$.

Claim. There exists $r > 0$ such that given x in F with $\|x\|_F < 1$, we can find y in E with $\|y\|_E < r$ and $\|x - y\|_F < \frac{1}{2}$. If the claim were false, we could find x_n in F such that

$$\|x_n\|_F < 1 \qquad \text{for all } n \tag{1}$$

and

$$\|x_n - y\|_F \geq \tfrac{1}{2} \qquad \text{for every } y \text{ in } E \text{ with } \|y\|_E < n. \tag{2}$$

Inequality (1) implies that $\{x_n\} \in \tilde{F}$. Since $\tilde{E}$ is dense in $\tilde{F}$, there exists $\{y_n\}$ in $\tilde{E}$ such that $\|\{x_n\} - \{y_n\}\|_{\tilde{F}} \leq \frac{1}{4}$, that is,

$$\|x_n - y_n\|_F \leq \tfrac{1}{4} \qquad \text{for all } n. \tag{3}$$

Since $\{y_n\} \in \tilde{E}$, there exists a constant K such that

$$\|y_n\|_E \leq K \qquad \text{for all } n. \tag{4}$$

Inequalities (3) and (4) contradict (2) when $n \geq K$.

This proves the claim. Now to prove $F \subset E$, it suffices to show that every x in F with $\|x\|_F < 1$ belongs to E. By the claim, for such x, we can find y_1 in E such that $\|x - y_1\|_F < \frac{1}{2}$ and $\|y_1\|_E < r$. Now $\|2x - 2y_1\|_F < 1$. Again by the claim, we can find z_2 in E such that $\|z_2\|_E < r$ and $\|2x - 2y_1 - z_2\|_F < \frac{1}{2}$. Let $y_2 = z_2/2$. Then $y_2 \in E$, $\|y_2\|_E < r/2$, and $\|x - (y_1 + y_2)\|_F < (\frac{1}{2})^2$. Continuing in this way, we get a sequence of elements y_n in E such that for all n,

$$\|y_n\|_E < r(\tfrac{1}{2})^{n-1} \tag{5}$$

and

$$\|x - (y_1 + \cdots + y_n)\|_F < (\tfrac{1}{2})^n. \tag{6}$$

Since E is complete, (5) implies that the series $\sum_{n=1}^{\infty} y_n$ converges in E to, say, y. But (6) implies that the series converges in F to x. Hence $x = y \in E$. $\qquad\qquad\qquad\qquad\qquad\qquad\qquad\qquad\qquad\qquad\qquad\square$

Definition 2.4.4 Let X be a compact Hausdorff space and A a real subspace of $C(X)$. Let $\|\cdot\|$ be a norm on A. For u in Re A, we define the *quotient norm of u* as

$$\||u\|| := \inf\{\|u + iv\| : u + iv \in A\}.$$

Lemma 2.4.5 Let X be a compact Hausdorff space and A be a real subspace of $C(X)$. Let $\|\cdot\|$ be a norm on A.
(i) If $(A, \|\cdot\|)$ is continuously injected in $C(X)$ (with the supremum norm), then (Re A, $\||\cdot\||$) is continuously injected in $C_{\mathbb{R}}(X)$. In particular, this happens if $\|\cdot\|$ is the supremum norm.
(ii) If $(A, \|\cdot\|)$ is a Banach space, then (Re A, $\||\cdot\||$) is a Banach space.
(iii) If $\|\cdot\|$ is the supremum norm, A is a Banach algebra and Re A is a ring, then there is a constant $K > 0$ such that $\||uv\|| \leq K\||u\|| \, \||v\||$ for all u, v in Re A.

Proof. Our proof follows closely the proof of similar statements about complex subspaces and subalgebras as given in Burckel (1972). Let $\|\cdot\|_\infty$ denote the supremum norm.

(i) There exists $M > 0$ such that $\|f\|_\infty \leq M\|f\|$ for all f in A. Now for all f in A with Re $f = u$, we have $\|u\|_\infty \leq \|f\|_\infty \leq M\|f\|$. Hence $\|u\|_\infty \leq M\||u\||$. This proves (i).

(ii) Clearly, (Re A, $\||\cdot\||$) is a normed linear space. Let $\sum u_n$ be an absolutely convergent series in Re A. Then $\sum \||u_n\||$ is finite. By Definition 2.4.4, for each n, there exists f_n in A such that Re $f_n = u_n$ and $\|f_n\| < \||u_n\|| + 1/n^2$. Then $\sum \|f_n\|$ is finite. Since A is complete, $\sum f_n$ converges to a function f in A. Hence $\sum u_n = $ Re $f \in$ Re A. Thus (Re A, $\||\cdot\||$) is a Banach space.

(iii) Let Re A be a ring. We fix v in Re A and define F_v: Re $A \to$ Re A by $F_v(u) := uv$ for u in Re A. Then F_v is linear. We show that the graph of F_v is closed with respect to the quotient norm. Let $u_n \in$ Re A be such that as $n \to \infty$, $u_n \to u \in$ Re A and $F_v(u_n) = vu_n \to w$; that is, $\||u_n - u\|| \to 0$ and $\||u_n v - w\|| \to 0$. Since $\|u_n - u\|_\infty \le \||u_n - u\||$ and $\|u_n v - w\|_\infty \le \||u_n v - w\||$, we have for all x in X, $u_n(x) \to u(x)$ and $u_n(x)v(x) \to w(x)$. Thus $uv = w = F_v(u)$, and F_v has a closed graph. Hence F_v is continuous by the closed graph theorem. Let $K(v)$ be the operator norm of F_v. Then

$$\||uv\|| \le K(v)\||u\|| \qquad \text{for all } u \text{ in Re } A.$$

Now for a fixed u, the family $\{F_v(u) : v \in \text{Re } A, \||v\|| \le 1\}$ is bounded, because $\||F_v(u)\|| = \||uv\|| \le K(u)\||v\|| \le K(u)$. Hence, by the uniform boundedness principle, $\{K(v) : v \in \text{Re } A, \||v\|| \le 1\}$ is bounded by a finite positive constant K. Hence $\||uv\|| \le \||F_v(u)\|| \le K(v)\||u\|| \le K\||u\||$ for all v in Re A with $\||v\|| \le 1$. Hence for all u, v in Re A, we have $\||uv\|| \le K\||u\||\,\||v\||$. $\qquad\square$

Corollary 2.4.6 Let X be a compact Hausdorff space, τ a topological involution on X and A a real function algebra on (X, τ). If Re A is a ring, then $(\text{Re } A, \||\cdot\||)^{\tilde{}}$ is also a ring.

Proof. Since $(\text{Re } A, \||\cdot\||)^{\tilde{}} := l^\infty(\mathbb{N}, (\text{Re } A, \||\cdot\||))$ is a linear space and $ab = \frac{1}{2}[(a + b)^2 - a^2 - b^2]$, it suffices to show that it is closed under squaring. Let $\{u_n\}$ be a bounded sequence in $(\text{Re } A, \||\cdot\||)$, $\||u_n\|| \le M$ for all n. Then by Theorem 2.4.5(iii), $\||u_n^2\|| \le K\||u_n\||^2 \le KM^2$ for all n; that is, $\{u_n^2\}$ is a bounded sequence in $(\text{Re } A, \||\cdot\||)$.

Remark 2.4.7 Recall that every completely regular space Y can be embedded as a dense subspace of a unique (up to homeomorphism) compact Hausdorff space $\beta(Y)$, known as the *Stone–Čech compactification* of Y [see Simmons (1963)]. Let X be a compact Hausdorff space and E a (real or complex) normed linear space continuously injected in $C(X)$. Then by Lemma 2.4.3, $\tilde{E}$ is continuously injected in $C(X)^{\tilde{}}$. Every element in $C(X)^{\tilde{}}$ is a bounded sequence of continuous functions $\{f_n\}$. We can identify this sequence with the continuous bounded function f on $\mathbb{N} \times X$ defined by $f(n, x) = f_n(x)$ for (n, x) in $\mathbb{N} \times X$. Then f has a unique continuous extension to $\beta(\mathbb{N} \times X)$. Thus we can identify $C(X)^{\tilde{}}$ with $C(\beta(\mathbb{N} \times X))$ and we can regard $\tilde{E}$ as a subset of $C(\beta(\mathbb{N} \times X))$. Now let τ be a topological involution on X. For (n, x) in $\mathbb{N} \times X$, we define $\tau_\mathbb{N}(n, x) = (n, \tau(x))$. Then $\tau_\mathbb{N}$ is a continuous involution on $\mathbb{N} \times X$ and it has a unique continuous extension, say, τ_β to $\beta(\mathbb{N} \times X)$. Obviously, τ_β is an involution on $\beta(\mathbb{N} \times X)$. As mentioned above, we can identify $C(X, \tau)^{\tilde{}}$

$:= l^\infty(\mathbb{N}, C(X,\tau))$ with $C(\beta(\mathbb{N} \times X), \tau_\beta)$ and $C_\mathbb{R}(X,\tau)\tilde{}$ with $C_\mathbb{R}(\beta(\mathbb{N} \times X), \tau_\beta)$. Hence if E is a real normed linear space, continuously injected in $C(X,\tau)$ [respectively, $C_\mathbb{R}(X,\tau)$], then $\tilde{E}$ can be regarded as a subset of $C(\beta(\mathbb{N} \times X), \tau_\beta)$ [respectively, $C_\mathbb{R}(\beta(\mathbb{N} \times X), \tau_\beta)$].

Lemma 2.4.8 Let X be a compact Hausdorff space, τ a topological involution on X and A a real function algebra on (X,τ). Suppose that Re A is dense in $C_\mathbb{R}(X,\tau)$. Then for every a,b in $\beta(\mathbb{N} \times X)$ with $b \notin \{a,\tau_\beta(a)\}$, there exists f in $\tilde{A}$ such that Re $f(a) \neq$ Re $f(b)$; that is, Re$(\tilde{A})$ separates the points of $\beta(\mathbb{N} \times X)/_{\tau_\beta}$.

Proof. As noted in Remark 2.4.7, $\tilde{A}$ can be regarded as a subalgebra of $C(\beta(\mathbb{N} \times X), \tau_\beta)$. Obviously, $\tilde{A}$ contains constants.

Claim. Let $\varepsilon > 0$ and $g \in C_\mathbb{R}(X,\tau)$ with $g > 0$. Then there exists h in A such that

$$\| g - |\exp(h)| \, \| < \varepsilon \exp(\varepsilon)\|g\|.$$

Let $u := \log g \in C_\mathbb{R}(X,\tau)$. Since Re A is dense in $C_\mathbb{R}(X,\tau)$, there exists h in A such that $\|u - \text{Re } h\| < \varepsilon$. We may apply the mean value theorem to $\exp(u(x)) - \exp(\text{Re } h(x))$ for each x in X and get

$$\| g - |\exp(h)| \, \| = \|\exp(u) - \exp(\text{Re } h)\|$$
$$\leq \|u - \text{Re } h\| \exp(\|u - \text{Re } h\| + \|u\|)$$
$$< \varepsilon \exp(\varepsilon)\|g\|,$$

as desired.

Now let a,b in $\beta(\mathbb{N} \times X)$ be such that $b \notin \{a,\tau_\beta(a)\}$. Clearly, $\{a,\tau_\beta(a)\}$ and $\{b,\tau_\beta(b)\}$ are disjoint closed subsets of $\beta(\mathbb{N} \times X)$. By Urysohn's lemma, there exists ϕ in $C(\beta(\mathbb{N} \times X))$, such that $\phi(\{a,\tau_\beta(a)\}) = \{4\}$, $\phi(\{b,\tau_\beta(b)\}) = \{1\}$, and $1 \leq \phi \leq 4$. Let $\psi(z) := (\phi(z) + \phi(\tau_\beta(z)))/2$ for z in $\beta(\mathbb{N} \times X)$. Then $\psi \in C(\beta(\mathbb{N} \times X),\tau_\beta)$, $\psi(a) = 4$, $\psi(b) = 1$, and $1 \leq \psi \leq 4$. Regarding ψ as $\{\psi_n\}$ in $l^\infty(\mathbb{N},C(X,\tau))$, we have each $\psi_n > 0$. Hence by the claim, there exists $g_n \in A$ such that

$$\| \psi_n - |\exp(g_n)| \, \| \leq 1.$$

Since $\{\psi_n\}$ is bounded, $\{p_n\} := \{\exp(g_n)\}$ is bounded. Hence $p := \{p_n\} \in \tilde{A} \subset C(\beta(\mathbb{N} \times X, \tau_\beta))$. From (i), since $\mathbb{N} \times X$ is dense in $\beta(\mathbb{N} \times X)$, we have $\| \psi - |p| \, \| \leq 1$. Hence $|4 - |p(a)|| = |\psi(a) - |p(a)|| \leq 1$ and $|1 - |p(b)|| = |\psi(b) - |p(b)|| \leq 1$. Thus $|p(a)| \neq |p(b)|$. Let $p(a) = x_1 + iy_1$ and $p(b) = x_2 + iy_2$. If $x_1 \neq x_2$, we take $f = p$. If $x_1 = x_2$, we must have $y_1^2 \neq y_2^2$. We take $f = -(p - x_1)^2 \in \tilde{A}$. Then Re $f(a) = y_1^2 \neq y_2^2 = $ Re $f(b)$. $\qquad\square$

Theorem 2.4.9 Let X be a compact Hausdorff space, τ a topological involution on X, and A a real function algebra on (X,τ). If Re A is a ring, then $A = C(X,\tau)$.

Proof. Let $\|\|\cdot\|\|$ denote the quotient norm on Re A. By Lemma 2.4.5, (Re A, $\|\|\cdot\|\|$) is a Banach space, continuously injected in $C_{\mathbb{R}}(X,\tau)$. Hence by Lemma 2.4.3, (Re A, $\|\|\cdot\|\|$)$^{\sim}$ is continuously injected in $C_{\mathbb{R}}(X,\tau)^{\sim}$, which can be identified with $C_{\mathbb{R}}(\beta(\mathbb{N} \times X), \tau_{\beta})$.

We shall show that (Re A, $\|\|\cdot\|\|$)$^{\sim}$ is dense in $C_{\mathbb{R}}(\beta(\mathbb{N} \times X), \tau_{\beta})$. By Lemma 1.3.9, Re A separates the points of X/τ. The fact that Re A is a subalgebra of $C_{\mathbb{R}}(X,\tau)$ containing 1 implies that Re A is uniformly dense in $C_{\mathbb{R}}(X,\tau)$ by Corollary 2.1.4. Hence Lemma 2.4.8 implies that Re$(\tilde{A})$ separates the points of $\beta(\mathbb{N} \times X)/\tau_{\beta}$. It is elementary to check that Re $\tilde{A} = ($Re A, $\|\|\cdot\|\|)^{\sim}$. Further, by Corollary 2.4.6, (Re A, $\|\|\cdot\|\|$)$^{\sim}$ is a ring. Thus (Re A, $\|\|\cdot\|\|$)$^{\sim}$ is a subalgebra of $C_{\mathbb{R}}(\beta(\mathbb{N} \times X), \tau_{\beta})$ containing 1. Hence, by Corollary 2.1.4, (Re A, $\|\|\cdot\|\|$)$^{\sim}$ is dense in $C_{\mathbb{R}}(\beta(\mathbb{N} \times X), \tau_{\beta}) = C_{\mathbb{R}}(X,\tau)^{\sim}$. Now Lemma 2.4.3 implies that Re $A = C_{\mathbb{R}}(X,\tau)$. Hence by Corollary 2.3.4, $A = C(X,\tau)$. $\square$

The above analog of Wermer's theorem is proved in Kulkarni and Srinivasan (1990). This proof is also given in Hwang (1990).

Remark 2.4.10 Let A be a real uniform algebra (Definition 1.3.13). We have shown in Theorem 1.3.17 that A can be regarded as a real function algebra on (X,τ), where $X = \mathrm{Car}(A)$ and $\tau(\phi) = \bar{\phi}$ for ϕ in X. Identifying the elements in A with their Gelfand transforms, we can denote this real function algebra by A itself. We have shown in Remark 1.2.11 that the maximal ideal space $M(A)$ of A is the quotient space obtained from X by identifying ϕ with $\tau(\phi)$ for each ϕ in X. Recall that for a in A, Re $\hat{a}(M) = $ Re $\phi(a) = $ Re $a(\phi)$ for M in $M(A)$ and ϕ in X with ker $\phi = M$. Thus Re $\hat{a} \circ \ker = $ Re a for each a in A. Hence if the subspace Re $\hat{A}$ of $C_{\mathbb{R}}(M(A))$ is a ring, then the subspace Re A of $C_{\mathbb{R}}(X,\tau)$ is a ring. But then by Theorem 2.4.9, $A = C(X,\tau)$, Re $A = C_{\mathbb{R}}(X,\tau)$, and hence Re $\hat{A} = C_{\mathbb{R}}(M(A))$. Similarly, the relation Re $\hat{a} \circ \ker = $ Re a also shows that $\|$Re $\hat{a}\|_{\infty} = \|$Re $a\|_{\infty}$ for each a in A. Hence if Re $\hat{A}$ is uniformly closed in $C_{\mathbb{R}}(M(A))$, then Re A is uniformly closed in $C_{\mathbb{R}}(X,\tau)$. Now, by Theorem 2.3.3, $A = C(X,\tau)$ and, as above, Re $\hat{A} = C_{\mathbb{R}}(M(A))$.

2.5 FUNCTIONS THAT OPERATE IN RE A

Definition 2.5.1 Let X be a compact Hausdorff space, E a subset of $C(X)$, J a subset of $\mathbb{R}$, and $h: J \to \mathbb{R}$.

(i) We say that h *operates in* E if for every f in E with $f(X) \subset J$, $h \circ f$ belongs to E.

(ii) Let E be a real subspace of $C(X)$ and $\|\cdot\|$ be a norm on E. We say that h *operates boundedly in* E if for every $\varepsilon > 0$, there exists $M > 0$ such that $h \circ f$ belongs to E and $\|h \circ f\| < M$ for every f in E with $f(X) \subset J$ and $\|f\| < \varepsilon$.

Let X be a compact Hausdorff space and A a complex function algebra on X. Since Re A is a real vector space containing the constant function 1, it is clear that every affine function on an interval operates in Re A.† Recall that a real-valued function h on an interval I is affine if $h(t) = at + b$ for some real numbers a and b and all t in I. It was conjectured in the early 1960s that if a nonaffine function operates in Re A, then $A = C(X)$. Wermer's theorem discussed in the preceding section proves the conjecture in the case $h(t) = t^2$ for t in $\mathbb{R}$. Similarly, a well-known theorem of Bernard (1974) states that if Re A is a lattice, then $A = C(X)$. This amounts to proving the conjecture in the case $h(t) = |t|$ for t in $\mathbb{R}$ since $\max(u,v) = \frac{1}{2}(u + v + |u - v|)$, $\min(u,v) = \frac{1}{2}(u + v - |u - v|)$, and $|u| = \max(u,0) - \min(u,0)$. A major step in proving the conjecture above was taken by Bernard in 1970. He showed that if some continuous nonaffine function on an interval operates boundedly in Re A with its quotient norm (Definition 2.4.4), then $A = C(X)$. In this section an analog of this theorem for real function algebras is proved.

We also prove an analog of a theorem due to Sidney (1979) which states that if h is continuous on an interval I, if h is nonaffine on every subinterval of I, and if h operates in Re A, then $A = C(X)$.

In the case of a complex function algebra A, Hatori (1981) proves that if a continuous nonaffine function on an interval operates in Re A, then $A = C(X)$. Jarosz and Sawon (1985) prove that a discontinuous function operates in Re A if and only if A is finite dimensional, and in that case, X is a finite set. As a consequence, it follows that if A is a complex function algebra and a nonaffine function operates in Re A, then $A = C(X)$, proving the original conjecture. A similar result in the case of a real function algebra is not yet available.

Lemma 2.5.2 Let h be a continuous function on $[-1,2]$, q a continuous function on $[-1,1]$, and let p be defined by

---

†By an interval we shall always mean a connected subset of $\mathbb{R}$ that does not reduce to a singleton.

$$p(t) := \int_{-1}^{1} h(s + t)q(s)\, ds, \qquad 0 \le t \le 1. \tag{1}$$

Let X be a compact Hausdorff space and E a uniformly closed subspace of $C_{\mathbb{R}}(X)$ containing 1. Let h operate in E. Then p also operates in E.

Proof. For $n = 1,2,\ldots$, let $\mathscr{P}_n := \{s_0,\ldots,s_{2^n}\}$ be the partition obtained by dividing $[-1,1]$ into 2^n equal subintervals:

$$s_j = -1 + j \cdot 2^{1-n}, \qquad j = 0,1,\ldots,2^n. \tag{2}$$

For $0 \le t \le 1$, let

$$p_n(t) := \frac{1}{2^{n-1}} \sum_{j=1}^{2^n} h(t + s_j)q(s_j). \tag{3}$$

Since E contains constants, each $p_n(t)$ operates in E. Further, since $p_n(t)$ is a Riemann sum of the integral in (1), $\{p_n(t)\}$ converges to $p(t)$ for $0 \le t \le 1$. We shall show that this convergence is uniform. Note that for $m = 1,2,\ldots$, the partition $\mathscr{P}_{n+m}$ can be obtained by further dividing each subinterval $[s_{j-1},s_j]$ into 2^m subintervals of equal length. Let

$$s_{j,k} = s_{j-1} + k(2^{1-n-m}), \qquad k = 0,1,\ldots,2^m, \quad j = 1,\ldots,2^n. \tag{4}$$

Thus we can write

$$p_{n+m}(t) = \frac{1}{2^{n+m-1}} \sum_{j=1}^{2^n} \sum_{k=1}^{2^m} h(t + s_{j,k})q(s_{j,k}). \tag{5}$$

From (3) and (5), we obtain

$$|p_{n+m}(t) - p_n(t)| \le \frac{1}{2^{n-1}} \sum_{j=1}^{2^n} \frac{1}{2^m} \sum_{k=1}^{2^m} |h(t + s_{j,k})q(s_{j,k}) - h(t + s_j)q(s_j)|. \tag{6}$$

Since h is continuous on $[-1,2]$ and q is continuous on $[-1,1]$, we can find $M > 0$ such that $|h(x)| < M$ for x in $[-1,2]$ and $|q(y)| < M$ for y in $[-1,1]$. Now let $\varepsilon > 0$ be given. There exists $\delta > 0$ such that if $x,y \in [-1,2]$ and $|x - y| < \delta$, then $|h(x) - h(y)| < \varepsilon/4M$ and $|q(x) - q(y)| < \varepsilon/4M$. It follows from (2) and (4) that

$$|s_{j,k} - s_j| < \frac{1}{2^{n-1}} \qquad \text{for } j = 1,\ldots,2^n, \quad k = 1,\ldots,2^m.$$

Now we can choose n_0 such that $1/2^{n_0-1} < \delta$. Then for every $n \geq n_0$ and $m = 1,2,\ldots$, we have

$$|h(t + s_{j,k})q(s_{j,k}) - h(t + s_j)q(s_j)|$$

$$\leq |h(t + s_{j,k}) - h(t + s_j)| \, |q(s_{j,k})| + |h(t + s_j)| \, |q(s_{j,k}) - q(s_j)|$$

$$< \frac{\varepsilon}{4M} M + M \frac{\varepsilon}{4M} = \frac{\varepsilon}{2} \qquad \text{for } 0 \leq t \leq 1. \tag{7}$$

From (6) and (7), we obtain $|p_{n+m}(t) - p_n(t)| < \varepsilon$ for $0 \leq t \leq 1$. Thus $\{p_n(t)\}$ is a uniformly Cauchy sequence and hence converges uniformly to $p(t)$ on $[0,1]$. Hence if $u \in E$ and $u(X) \subset [0,1]$, then $\{p_n \circ u\}$ converges to $p \circ u$ uniformly. But $p_n \circ u \in E$ for each n as p_n operates in E and E is uniformly closed. Hence $p \circ u \in E$. Thus p operates in E. $\qquad \square$

Lemma 2.5.3 Let X be a compact Hausdorff space and E be a subspace of $C_{\mathbb{R}}(X)$ containing 1. If a continuous nonaffine function h on an interval I operates in E, then the uniform closure of E is a ring.

Proof. The proof depends on showing that a polynomial of degree 2 defined on $[0,1]$ operates in the uniform closure of E, say, F.

Since h is nonaffine on I, there exist a,b in I with $a < b$ and t_0 in $(0,1)$ such that

$$h((1 - t_0)a + t_0 b) \neq (1 - t_0)h(a) + t_0 h(b). \tag{1}$$

For $0 \leq t \leq 1$, we define $p(t)$ by

$$p(t) := h((1 - t)a + tb) - (1 - t)h(a) - th(b). \tag{2}$$

Then p is a continuous function on $[0,1]$, $p(0) = 0 = p(1)$, whereas from (1), $p(t_0) \neq 0$. Thus p is nonaffine on $[0,1]$. Clearly, p operates in E. We shall now show that p operates in F.

Let $u \in F$ and $u(X) \subset [0,1]$. For each $n = 1,2,\ldots$, there exists u_n in E such that $\|u_n - u\|_\infty < 1/n$. Then $u_n(X) \subset [-1/n, 1 + 1/n]$. Let $v_n := (u_n + 1/n)/(1 + 2/n)$. Then $v_n \in E$ and $v_n(X) \subset [0,1]$. Hence $p \circ v_n \in E$ for each $n = 1,2,\ldots$. Further, p is continuous and $v_n \to u$ uniformly as $n \to \infty$. Hence $p \circ v_n \to p \circ u$ uniformly as $n \to \infty$. Therefore, $p \circ u \in F$. Thus p operates in F.

Now we define p to be zero outside $[0,1]$, so that p is uniformly continuous on the real line. For $n = 1,2,\ldots$, we define

$$q_n(t) := c_n(1 - t^2)^n, \qquad -1 \leq t \leq 1, \tag{3}$$

where

$$c_n^{-1} := \int_{-1}^{1} (1 - t^2)^n \, dt, \tag{4}$$

and

$$p_n(t) := \int_{-1}^{1} p(t + s) q_n(s) \, ds, \qquad -0 \le t \le 1. \tag{5}$$

By Lemma 2.5.2, $p_n(t)$ operates in F for $n = 1, 2, \ldots$. It is well known that the sequence $\{p_n\}$ defined as above converges to p uniformly, as in the classical proof of Weierstrass's theorem [see, e.g., Rudin (1964, p. 147)]. Let

$$0 < \varepsilon < \frac{|p(t_0)|}{2}. \tag{6}$$

When n is sufficiently large, $|p_n(t) - p(t)| < \varepsilon$ for $0 \le t \le 1$. For such n, we have $|p_n(0)| < \varepsilon$, $|p_n(1)| < \varepsilon$ because $p(0) = 0 = p(1)$, and $|p_n(t_0)| > \varepsilon$ because of (6). Thus p_n is nonaffine on $[0,1]$.

A simple change of variable $y = t + s$ in (5) implies that

$$p_n(t) := \int_{-1+t}^{1+t} p(y) q_n(y - t) \, dy = \int_{0}^{1} p(y) q_n(y - t) \, dy \qquad \text{for } 0 \le t \le 1,$$

as p is 0 outside $[0,1]$. This shows that p_n is a polynomial on $[0,1]$. Since p_n is nonaffine, the degree m of p_n is at least 2. If $m = 2$, then the polynomial p_n of degree 2 defined on $[0,1]$ operates in F. If $m > 2$, then $p_n((1 + t)/2) - p_n(t/2), 0 \le t \le 1$ is a polynomial of degree $(m - 1)$ on $[0,1]$ and it operates in F.

Proceeding in this way, we get a polynomial of degree 2 defined on $[0,1]$ that operates in F. Hence the function $f(t) = t^2, 0 \le t \le 1$, operates in F. Now let $u \in F$. We can find $M > 0$ such that $u(X) \subset [-M, M]$. Then $w := (u + M)/2M \in F$ and $w(X) \subset [0,1]$. Hence $f \circ w = w^2 \in F$ and $u^2 = (2Mw - M)^2 \in F$. This shows that F is a ring. $\qquad \square$

Lemma 2.5.4 Let X be a compact Hausdorff space and E a subspace of $C_{\mathbb{R}}(X)$ that contains 1. Suppose that a continuous nonaffine function on an interval I operates in E.
(i) If E separates the points of X, then E is uniformly dense in $C_{\mathbb{R}}(X)$.
(ii) Let τ be an involution on X and E be a subspace of $C_{\mathbb{R}}(X, \tau)$ separating the points of X/τ. Then E is uniformly dense in $C_{\mathbb{R}}(X, \tau)$.

Proof. Let F be the uniform closure of E. Then F is a ring by Lemma 2.5.3. Now (ii) follows from Corollary 2.1.4 and (i) follows from the comments below Corollary 2.1.4 (by the real Stone–Weierstrass theorem). $\qquad\square$

Statement (i) of Lemma 2.5.4 was proved in deLeeuw and Katznelson (1963). Burckel (1972) also contains a proof.

Theorem 2.5.5 Let X be a compact Hausdorff space, τ a topological involution on X, and A a real function algebra on (X,τ). If some continuous nonaffine function h on an interval I operates boundedly in Re A with its quotient norm, then $A = C(X,\tau)$.

Proof. Let $\|\!\|\!\|\cdot\|\!\|\!\|$ denote the quotient norm on Re A. By Lemma 2.4.5, (Re A, $\|\!\|\!\|\cdot\|\!\|\!\|$) is a Banach space continuously injected in $C_{\mathbb{R}}(X,\tau)$. Hence by Lemma 2.4.3, (Re A, $\|\!\|\!\|\cdot\|\!\|\!\|)^{\sim}$ is continuously injected in $C_{\mathbb{R}}(X,\tau)^{\sim}$, which can be identified with $C_{\mathbb{R}}(\beta(\mathbb{N} \times X), \tau_{\beta})$ as in Remark 2.4.7. We shall show that (Re A, $\|\!\|\!\|\cdot\|\!\|\!\|)^{\sim}$ is dense in $C_{\mathbb{R}}(\beta(\mathbb{N} \times X), \tau_{\beta})$.

First, by Lemma 1.3.9, Re A separates the points of X/τ. Since a continuous nonaffine function h operates in Re A, Lemma 2.5.4 implies that Re A is uniformly dense in $C_{\mathbb{R}}(X,\tau)$. Hence by Lemma 2.4.8, Re $\tilde{A}$ separates the points of $\beta(\mathbb{N} \times X)/\tau_{\beta}$. Suppose that $u = \{u_n\}$ is in (Re A, $\|\!\|\!\|\cdot\|\!\|\!\|)^{\sim}$ and $u_n(X) \subset I$ for all n. Then $\|\!\|\!\|u_n\|\!\|\!\| < \varepsilon$ for some $\varepsilon > 0$ and for all $n = 1,2,\dots$. Since h operates boundedly in (Re A, $\|\!\|\!\|\cdot\|\!\|\!\|$), there exists $M > 0$ such that $h \circ u_n$ belongs to Re A and $\|\!\|\!\|h \circ u_n\|\!\|\!\| < M$ for all $n = 1,2,\dots$. Thus h operates in (Re A, $\|\!\|\!\|\cdot\|\!\|\!\|)^{\sim}$. Hence again by Lemma 2.5.4, (Re A, $\|\!\|\!\|\cdot\|\!\|\!\|)^{\sim}$ is uniformly dense in $C_{\mathbb{R}}(\beta(\mathbb{N} \times X), \tau_{\beta}) = C_{\mathbb{R}}(X,\tau)^{\sim}$. Now Lemma 2.4.3 implies that Re $A = C_{\mathbb{R}}(X,\tau)$, and hence by Corollary 2.3.4, $A = C(X,\tau)$. $\qquad\square$

Theorem 2.5.6 (Hwang, 1990) Let X be a compact Hausdorff space, τ a topological involution on X, and A a real function algebra on (X,τ). Suppose that h is a continuous real-valued function on an interval I such that h is not affine on any subinterval of I. If h operates in Re A, then $A = C(X,\tau)$.

Proof. As noted in the proof of Theorem 2.5.5, (Re A, $\|\!\|\!\|\cdot\|\!\|\!\|)^{\sim}$ is a subspace of $C_{\mathbb{R}}(\beta(\mathbb{N} \times X), \tau_{\beta})$. Let $\tilde{E}$ denote the uniform closure of (Re A, $\|\!\|\!\|\cdot\|\!\|\!\|)^{\sim}$ in $C_{\mathbb{R}}(\beta(\mathbb{N} \times X), \tau_{\beta})$, and let

$$\tilde{F} := \{\tilde{u} \in C_{\mathbb{R}}(\beta(\mathbb{N} \times X), \tau_{\beta}) : \tilde{u}\tilde{w} \in \tilde{E} \text{ for all } \tilde{w} \in \tilde{E}\}.$$

Then $\tilde{F}$ is a uniformly closed subalgebra of $C_{\mathbb{R}}(\beta(\mathbb{N} \times X), \tau_{\beta})$, contains 1, and is contained in $\tilde{E}$. We shall prove that $\tilde{F}$ separates the points of $\beta(\mathbb{N} \times X)/\tau_\beta$. Then by Corollary 2.1.4, $\tilde{F} = C_\mathbb{R}(\beta(\mathbb{N} \times X), \tau_\beta)$, and hence $\tilde{E} = C_\mathbb{R}(\beta(\mathbb{N} \times X), \tau_\beta)$; that is, (Re A, $\||\cdot\||)^\sim$ is dense in $C_\mathbb{R}(X,\tau)^\sim$. Now by Lemma 2.4.3, Re $A = C_\mathbb{R}(X,\tau)$, so that by Corollary 2.3.4, $A = C(X,\tau)$.

To complete the proof, we show that $\tilde{F}$ separates the points of $\beta(\mathbb{N} \times X)/\tau_\beta$. Choose $a < b$ such that $[a,b] \subset I$. Let $D := \{u \in \text{Re } A : a \le u \le b\}$, and for $n = 1,2,\ldots$, let $D_n := \{u \in D : \||h \circ u\|| < n\}$. D is closed in (Re A, $\||\cdot\||)$, which is a Banach space by Lemma 2.4.5, so that D is complete. Since $D = \cup_n D_n$, the closure of some D_n has a nonempty interior in D by the Baire category theorem. Thus there are $u_0 \in D$, $\eta > 0$ and a natural number m such that $U := \{u \in \text{Re } A : \||u - u_0\|| < 3\eta\}$ is contained in the closure of D_m.

If necessary, we may replace u_0 by $su_0 + t$ for suitable real numbers s and t, $0 < s < 1$, and shrink η to arrange that $U \subset D$ and

$$U \cap D_m \text{ is dense in } U. \tag{1}$$

Let $\bar{u}_0$ denote the sequence $u_0, u_0, \ldots$. For a sequence $\bar{u} = \{u_n\}$ in (Re A, $\||\cdot\||)^\sim$, recall that $\|\bar{u}\|_\infty = \sup\{\||u_n\|| : n = 1,2,\ldots\}$. For $\varepsilon > 0$, let

$$\tilde{W}_\varepsilon := \{\bar{u} \in (\text{Re } A, \||\cdot\||)^\sim : \|\bar{u} - \bar{u}_0\|_\infty < \varepsilon\}. \tag{2}$$

Claim 1. $h \circ \bar{u}$ belongs to $\tilde{E}$ for every $\bar{u}$ in $\tilde{W}_{3\eta}$.

Let $\bar{u} \in \tilde{W}_{3\eta}$ and let $\bar{w} = \bar{u} - \bar{u}_0 = \{w_n\}$, so that $\||\bar{w}\|| < 3\eta$, $u_0 + w_n \in U$ for each n. By (1) for natural numbers n and k, we can choose $u(n,k)$ in $U \cap D_m$ such that $\||u(n,k) - (u_0 + w_n)\|| < 1/k$. For each fixed k, let $\bar{u}_k = \{u(n,k)\} \in (\text{Re } A, \||\cdot\||)^\sim$. Since $u(n,k) \in D_m$, $\||h \circ u(n,k)\|| < m$, and hence $h \circ \bar{u}_k = \{h \circ u(n,k)\}$ belongs to (Re A, $\||\cdot\||)^\sim$. As $k \to \infty$, $h \circ \bar{u}_k$ converges uniformly on $\beta(\mathbb{N} \times X)$ to $h \circ (\bar{u}_0 + \bar{w}) = h \circ \bar{u}$. Thus $h \circ \bar{u} \in \tilde{E}$.

For $0 < \delta < \eta$, let λ_δ be a nonnegative continuously differentiable function in $\mathbb{R}$ with support in $(-\delta, \delta)$ and satisfying $\int_{-\delta}^\delta \lambda_\delta(t)\,dt = 1$. Let h_δ be the convolution of h and λ_δ:

$$h_\delta(x) = \int_{-\delta}^\delta h(x - t)\lambda_\delta(t)\,dt = \int_a^b h(t)\lambda_\delta(x - t)\,dt.$$

Then h_δ is continuously differentiable on a neighborhood of $[a + \eta, b - \eta]$ and as $\delta \to 0$, $h_\delta \to h$ uniformly on $[a + \eta, b - \eta]$ (Burckel, 1972, Lemmas 4.19 and 4.20).

Claim 2. $h'_\delta \circ \bar{u}$ belongs to $\tilde{F}$ for every $\bar{u}$ in $\tilde{W}_{2\eta}$.

Let $\bar{u} \in \tilde{W}_{2\eta}$. If $t \in [-\delta, \delta]$, then by (2) $\bar{u} - t \in \tilde{W}_{3\eta}$, so that $h \circ (\bar{u} - t) \in \tilde{E}$, by claim 1. Since $\tilde{E}$ is uniformly closed, it can be proved (as in

Lemma 2.5.2) that $h_\delta \circ \bar{u} = \int h(\bar{u} - t)\lambda_\delta(t)\, dt$ belongs to $\tilde{E}$. Now if $\tilde{w} \in$ (Re A, $\|\|\cdot\|\|)^\sim$, then $\bar{u} + t\tilde{w} \in \tilde{W}_{2\eta}$ for small nonzero t. Hence $(h_\delta \circ (\bar{u} + t\tilde{w}) - h_\delta(\bar{u}))/t \in \tilde{E}$, and letting $t \to 0$, $(h_\delta' \circ \bar{u})\tilde{w} \in \tilde{E}$. Thus $h_\delta' \circ \bar{u} \in \tilde{F}$.

Now let $p, q \in \beta(\mathbb{N} \times X)$, $p \neq q$, $p \neq \tau_\beta(q)$. By Lemma 1.3.9, Re A separates the points of X/τ. Then by Lemma 2.5.4(ii), Re A is dense in $C_\mathbb{R}(X, \tau)$. Next, Lemma 2.4.8 implies that (Re A, $\|\|\cdot\|\|)^\sim$ separates the points of $\beta(\mathbb{N} \times X)/\tau_\beta$. We can choose $\tilde{w} \in \tilde{W}_\eta$ such that $\tilde{w}(p) \neq \tilde{w}(q)$. Let $\varepsilon > 0$ be such that $\varepsilon < \eta|\tilde{w}(p) - \tilde{w}(q)|/(2\|\tilde{w}\|_\infty)$. Then $(\tilde{w}(p) - \varepsilon, \tilde{w}(p) + \varepsilon) \subset I$. Hence we can choose t_1, t_2, and t_3 in $(\tilde{w}(p) - \varepsilon, \tilde{w}(p) + \varepsilon)$ in such a way that for $j = 1, 2, 3$, $(t_j, h(t_j))$ are not collinear. Since h_δ converges to h uniformly, we can choose a positive $\delta < \eta$ small enough so that the points $(t_j, h_\delta(t_j))$ are also not collinear. This ensures that h_δ' is not constant on $(\tilde{w}(p) - \varepsilon, \tilde{w}(p) + \varepsilon)$.

If $h_\delta'(\tilde{w}(p)) \neq h_\delta'(\tilde{w}(q))$, then let $\bar{u} = \tilde{w}$. If $h_\delta'(\tilde{w}(p)) = h_\delta'(\tilde{w}(q))$, then choose s in $(\tilde{w}(p) - \varepsilon, \tilde{w}(p) + \varepsilon)$ for which $h_\delta'(s) \neq h_\delta'(\tilde{w}(q))$, and let

$$\bar{u} = \tilde{w} + \frac{(\tilde{w} - \tilde{w}(q))(s - \tilde{w}(p))}{\tilde{w}(p) - \tilde{w}(q)}$$

so that $\|\bar{u} - \tilde{w}\|_\infty < \eta$, $\bar{u}(q) = \tilde{w}(q)$ and $\bar{u}(p) = s$. In either case, $\bar{u} \in \tilde{W}_{2\eta}$ and $h_\delta'(\bar{u}(p)) \neq h_\delta'(\bar{u}(q))$. Since $h_\delta' \circ \bar{u} \in \tilde{F}$ by claim 2, we see that $\tilde{F}$ separates the points of $\beta(\mathbb{N} \times X)/\tau_\beta$. $\square$

Remark 2.5.7 Let X be a compact Hausdorff space with a topological involution τ, and A be a real function algebra on (X, τ). Suppose that Re A is a ring. Then the function $h(t) = t^2$, $t \in \mathbb{R}$ operates in Re A: h is clearly nonaffine. It follows from Lemma 2.4.5 that there is a constant $K > 0$ such that $\|\|u^2\|\| \leq K\|\|u\|\|^2$ for all u in Re A. This shows that h operates boundedly in Re A with its quotient norm $\|\|\cdot\|\|$. Also, h is nonaffine on every subinterval of I. Thus Theorem 2.4.9 is a special case of Theorems 2.5.5 and 2.5.6.

Finally, we give an application of the results in this chapter to a class of real uniform algebras studied by Mehta and Vasavada (1986). As particular cases, we get the corresponding well-known theorems for complex function algebras.

Theorem 2.5.8 Let X be a compact Hausdorff space and A be a uniformly closed real subalgebra of $C(X)$ that contains real constants and separates the points of X. Suppose that A satisfies one of the following hypotheses:
(i) $\bar{f} \in A$ whenever $f \in A$.
(ii) Re A is uniformly closed.

(iii) Re A is a ring.

(iv) Some continuous nonaffine function on an interval I operates boundedly in Re A with its quotient norm.

(v) A continuous function h on an interval I which is nonaffine on every subinterval of I operates in Re A.

For $\phi \in \mathrm{Car}(A)$, let $\tau(\phi) = \bar{\phi}$. Identifying $x \in X$ with $e_x \in \mathrm{Car}(A)$, let $Y = X \cup \tau(X) \subset \mathrm{Car}(A)$. Then $\hat{A}_{|Y} = C(Y,\tau_{|Y})$.

Proof.   Immediate from Corollary 2.1.14 and Theorems 2.3.3, 2.4.9, 2.5.5, and 2.5.6, since A is isometrically isomorphic to $\hat{A}_{|Y}$ by Corollary 1.3.18.   □

Remark   2.5.9   In Theorem 2.5.8, suppose that A has one of the properties discussed in Remark 1.3.19 [in addition to satisfying one of the hypotheses (i) to (v)]:

(i) For each x in X, there is y in X such that $e_y = \bar{e}_x$. In this case, $Y = X$, so that $A = C(X,\tau)$.

(ii) For every x,y in X, $\bar{e}_x \neq e_y$. In this case, Y is a disjoint union of X and $\tau(X)$. Hence every f in $C(X)$ can be uniquely extended to a function in $C(Y,\tau)$ and we have $A = C(Y,\tau)|_X = C(X)$. We have shown in Remark 1.3.19 that every complex function algebra satisfies (ii). Thus we get the classical theorems about complex function algebras, namely, the Stone–Weierstrass theorem, the Hoffman–Wermer theorem, Wermer's theorem, Bernard's theorem, and Sidney's theorem as particular cases.

(iii) For every distinct x,y in X, $\bar{e}_x \neq e_y$. In this case, Y is a disjoint union of X and $\tau(X\setminus Z)$, where $Z := \{x \in X : f(x) \text{ is real for all } f \text{ in } A\}$. If $g \in C(Y,\tau)$, then g is real on Z, because $\tau(x) = x$ for x in Z. On the other hand, if $f \in C(X)$ is real on Z, then f has a unique extension to a function g in $C(Y,\tau)$ defined as follows:

$$g(x) := \begin{cases} f(x) & \text{for } x \text{ in } X \\ \bar{f}(\tau(x)) & \text{for } x \text{ in } \tau(X\setminus Z). \end{cases}$$

Then we have

$$A = C(Y,\tau)_{|X} = \{f \in C(X) : f \text{ is real on } Z\}.$$

In particular, Re $A = C_{\mathbb{R}}(X)$, which is the result of Mehta and Vasavada (1986).

Remark   2.5.10   Let X be a compact Hausdorff space, τ a topological involution on X, and A a real function algebra on (X,τ). Srinivasan (1988) has used the complexification technique to show that if A separates the points of $\beta(\mathbb{N} \times X)$ and if Re A is a lattice, then $A = C(X,\tau)$. It is not known whether the hypothesis of point separation can be dropped.

Employing the complexification technique along with the known theorems for complex function algebras, Srinivasan and Kulkarni (1988) show that $A = C(X,\tau)$ if any one of the following conditions is satisfied.

1. There exists a sequence $\{F_n\}$ of closed sets such that $X = \cup\,_n F_n$, $\tau(F_n) = F_n$, and $A_{|F_n} = C(F_n, \tau_{|F_n})$ for all n.
2. X does not have any nonempty perfect set.
3. Each x in X, with at most finitely many exceptions, has a compact neighborhood F_x such that $\tau(F_x) = F_x$ and $A_{|F_x} = C(F_x, \tau_{|F_x})$.

3

Gleason Parts of a Real Function Algebra

In this chapter we discuss the problem of embedding an analytic structure in the carrier space, and a harmonic structure in the carrier space or in the maximal ideal space of a real function algebra. These concepts are defined precisely in Section 3.1 and the relationship between the two is studied. A simplified statement of the problem of embedding an analytic structure is as follows.

If A is a real function algebra, then to what extent do (the Gelfand transforms of) the functions in A behave like analytic functions? In other words, how far is the real disk algebra of Example 1.3.10 prototypical? Clearly, if functions in A behave like analytic functions, their real parts behave like harmonic functions. A partial converse of this is proved by using an elementary result that if the real parts of the first four powers of a function are harmonic and if its real and imaginary parts have continuous partial derivatives up to the second order, then the function or its complex conjugate is analytic.

Gleason (1957) proved that for a complex function algebra A, the relation $\sim$ defined on $\mathrm{Car}(A)$ by $\phi \sim \psi$ "if and only if $\|\phi - \psi\| < 2$ (ϕ, ψ in $\mathrm{Car}(A)$)" is an equivalence relation. He proposed that the search for an analytic structure should proceed via the investigation of the metric topology on $\mathrm{Car}(A)$. This suggestion turned out to be fruitful in later years. A brief history of this problem can be found in Browder (1969) and Gamelin (1969).

In Section 3.2 we define a similar equivalence relation on the carrier space of a real function algebra A and give some characterizations of the equivalence classes under this relation. They are called *Gleason parts of A*. In Section 3.3 we study the relationship between the Gleason parts of A and those of its complexification B. A topological characterization of a Gleason part of A is given in Section 3.4. Section 3.5 gives some sufficient conditions for the existence of analytic and harmonic structures in the carrier space of A.

3.1 ANALYTIC AND HARMONIC STRUCTURE

Definition 3.1.1 Let G be a nonempty open subset of $\mathbb{C}$ and f a complex-valued function on G such that $\operatorname{Re} f$ and $\operatorname{Im} f$ have continuous partial derivatives of order k. Such functions will be referred to as C^k-*maps*. Recall the usual notations:

$$\frac{\partial}{\partial z} = \frac{1}{2}\left(\frac{\partial}{\partial x} - i\frac{\partial}{\partial y}\right), \qquad \frac{\partial}{\partial \bar{z}} = \frac{1}{2}\left(\frac{\partial}{\partial x} + i\frac{\partial}{\partial y}\right).$$

A function f is said to be *analytic* (respectively, *antianalytic*) on G if $\partial f/\partial \bar{z} = 0$ (respectively, $\partial f/\partial z = 0$) on G. It is called *dianalytic* on G if it is analytic or antianalytic on each connected component of G. A function u defined on G is said to be *harmonic* on G if u is a C^2-map and $u_{xx} + u_{yy} = 0$. (As usual, the subscripts denote the partial derivatives.) We may note that the real and the imaginary parts of analytic, antianalytic, or dianalytic functions on G are harmonic.

It can be easily proved that f is analytic if and only if $\bar{f}$ is antianalytic. [See Alling and Greenleaf (1971) for a proof.] Alling and Greenleaf (1971) also give a proof of the following lemma, which in effect says that every locally dianalytic function is dianalytic. We omit this proof, as the lemma is not used subsequently.

Lemma 3.1.2 Let f be a complex-valued function defined on a connected open subset G of the complex plane $\mathbb{C}$, whose real and imaginary parts u and v are C^2-maps. Assume that for all α in G, $(\partial f/\partial z)(\alpha) = 0$ or $(\partial f/\partial \bar{z})(\alpha) = 0$. Then f is analytic on G or antianalytic on G, so that u and v are harmonic on G.

Theorem 3.1.3 Suppose that $f = u + iv$ is analytic on a connected open subset G of $\mathbb{C}$, and w is a real-valued function on G. Let $g = u + iw$. Then:

(i) If u is nonconstant, then Re g^2 and Re g^3 are harmonic if and only if $w^2 = v^2 + av + b$, where a and b are constants.

(ii) Re g^2, Re g^3, and Re g^4 are harmonic if and only if $w^2 = (v + c)^2$ for some constant c.

(iii) Re g^2, Re g^3, and Re g^4 are harmonic and w is a C^1-map if and only if $w = v + c$ or $w = -(v + c)$ for some constant c. In this case g differs from f or $\bar{f}$ by a constant, and hence is analytic or antianalytic on G.

Proof. Let

$$p := \operatorname{Re} g^2 = u^2 - w^2, \tag{1}$$

$$q := \operatorname{Re} g^3 = u^3 - 3uw^2, \tag{2}$$

and

$$r := \operatorname{Re} g^4 = u^4 - 6u^2w^2 + w^4. \tag{3}$$

If $w^2 = v^2 + av + b$ for some constants a and b, then

$$p = u^2 - v^2 - av - b = \operatorname{Re} f^2 - av - b$$

and

$$q = u^3 - 3u(v^2 + av + b) = \operatorname{Re} f^3 - \frac{3a}{2} \operatorname{Im} f^2 - 3bu$$

are clearly harmonic. This proves the "if" part of (i). Similarly, the "if" parts of (ii) and (iii) can be proved.

To prove the "only if" parts, first assume that u is nonconstant. Let

$$G_1 := G \backslash \{z \in G : f'(z) = 0\}.$$

Since f is analytic and nonconstant, the zeros of f' are isolated. Hence G_1 is connected and is dense in G. Further, f is a conformal mapping on G_1. Now we can make a change of variables, $(x,y) \rightarrow (u,v)$ and regard p and q as harmonic functions of u and v on G_1, as a harmonic function remains harmonic under a change of variables arising from a conformal mapping. It follows from (1) that w^2 is a C^2-map as a function of u and v. Further from (1) and (2), we obtain

$$p_{uu} + p_{vv} = 2 - (w^2)_{uu} - (w^2)_{vv} = 0 \tag{4}$$

and

$$q_{uu} + q_{vv} = 3u[2 - (w^2)_{uu} - (w^2)_{vv}] - 6(w^2)_u = 0. \tag{5}$$

Equations (4) and (5) imply that $(w^2)_u = 0$; that is, w^2 is a function of v only. Hence $w^2 - v^2$ is also a function of v only. But $w^2 - v^2 = u^2 - v^2 - p = \operatorname{Re} f^2 - p$ is harmonic. Hence we must have $w^2 - v^2 = av +$

b for some constants a and b. Thus $w^2 = v^2 + av + b$ on G_1 and hence on G, because v and w are continuous on G and G_1 is dense in G. This proves the "only if" part of (i).

Now let r also be harmonic. Substituting $w^2 = v^2 + av + b$ in (3), we get $r_{uu} + r_{vv} = 2(a^2 - 4b) = 0$. Hence $b = a^2/4$, so that $w^2 = (v + a/2)^2$. If w is a C^1-map, then we have $w = v + a/2$ on G or $w = -v - a/2$ on G.

Next, assume that u is a constant. Then v is also constant. If p and r are harmonic, then (1) and (3) imply that w^2 and w^4 are harmonic. Hence w^2 is a constant. Thus we get the "only if" parts of (ii) and (iii).

The last statement of the theorem is obvious. $\qquad\square$

Remark 3.1.4
(a) For (i) and (ii) in Theorem 3.1.3, there is no regularity condition on w. In (i), if u is constant, we can only say that Re g^2 is harmonic if and only if w^2 is harmonic.
(b) For (iii), only the continuity of w is not enough, as the examples $g(z) = x + i|y|$ and $h(z) = x - i|y|$ show. Moreover, (ii) says that these are essentially (i.e., up to a change of variables) the only continuous counterexamples.

Corollary 3.1.5 Let $g = u + iw$ be a function on a connected open subset G of $\mathbb{C}$ such that Re g, Re g^2, Re g^3, and Re g^4 are harmonic. Then u has a uniquely determined (single-valued) harmonic conjugate u^* on G and $g = u \pm iu^*$ on G. Further, if w is a C^1-map, then g is analytic or antianalytic on G.

Proof. On every disk D lying in G, we consider a harmonic conjugate of u. Now (ii) of Theorem 3.1.3 says that $w^2 = (u^*)^2$ for a uniquely determined harmonic conjugate u^* of u on D. (The uniqueness of u^* is obvious.) Further, if u_1^* and u_2^* are harmonic conjugates of u on D_1 and D_2, respectively, satisfying $w^2 = (u_j^*)^2$ on $D_j, j = 1,2$, and $D_1 \cap D_2 \neq \varnothing$, then $u_1^* = \pm u_2^*$ on $D_1 \cap D_2$. Since u_1^* and u_2^* are both harmonic conjugates of u on $D_1 \cap D_2$ and $u_1^* = \pm u_2^*$, we must, in fact, have $u_1^* = u_2^*$. Hence u has a uniquely determined harmonic conjugate u^* on G such that $g = u \pm iu^*$ on G. Further, if g (i.e., w) is a C^1-map, then $g = u + iu^*$ on G or $g = u - iu^*$ on G. Thus g is analytic or antianalytic on G. $\qquad\square$

Lemma 3.1.6 Let G be an open connected set in $\mathbb{C}$ and A be a set of complex-valued continuous functions on G. Suppose that A is closed under multiplication, and every function in A is analytic or anitanalytic on

G. Then all functions in A are analytic, or all functions in A are antianalytic on G.

Proof. If possible, let f in A be analytic but not antianalytic, and g in A be antianalytic but not analytic. Since fg belongs to A, it is either analytic or antianalytic. If fg is analytic, then

$$0 = \frac{\partial(fg)}{\partial\bar{z}} = f\frac{\partial g}{\partial\bar{z}} + \frac{\partial f}{\partial\bar{z}}g = f\frac{\partial g}{\partial\bar{z}},$$

since f is analytic. Also, the analyticity of f implies that the zeros of f are isolated. Hence $\partial g/\partial\bar{z} = 0$ on G; that is, g is analytic on G, contrary to our assumption. Similarly, if fg is antianalytic, then we arrive at a contradiction by showing that f must be antianalytic. Hence the result. □

Lemma 3.1.7 Let V be an open set in $\mathbb{R}^n$ and A be a set of continuous complex-valued functions on V. Suppose that A is closed under multiplication and $\mathrm{Re}\,f$ is a C^k-map for each f in A. Let $Y := \{x \in V: \mathrm{Im}\,g(x) = 0$ for all g in $A\}$. Then $\mathrm{Im}\,f$ is a C^k-map on $V\backslash Y$ for each f in A.

Proof. Let $f = u + iv \in A$ and $x \in V\backslash Y$. Then there exists $g = u_1 + iv_1$ in A such that $v_1(x) \neq 0$. Since v_1 is continuous on V, one can find a neighborhood D of x such that $v_1(y) \neq 0$ for all y in D. We shall first show that v_1 is a C^k-map on D. $u_1 = \mathrm{Re}\,g$ and $u_1^2 - v_1^2 = \mathrm{Re}\,g^2$ are C^k-maps. Hence v_1^2 is a C^k-map. Let $w := v_1^2$. Since v_1 is nonzero on D, it is easy to check that v_1 has first partial derivatives on D and

$$\frac{\partial v_1}{\partial x} = \frac{1}{2v_1}\frac{\partial w}{\partial x}, \qquad \frac{\partial v_1}{\partial y} = \frac{1}{2v_1}\frac{\partial w}{\partial y}.$$

Similarly, we can show that v_1 has all partial derivatives up to order k, so that v_1 is a C^k-map on D.

By hypothesis, $u = \mathrm{Re}\,f$, $u_1 = \mathrm{Re}\,g$, and $uu_1 - vv_1 = \mathrm{Re}(fg)$ are C^k-maps. Hence vv_1 is a C^k-map. But on D, $v = vv_1/v_1$. Both vv_1 and v_1 are C^k-maps on D and v_1 is nonzero on D. Hence v is a C^k-map on D, and in particular at x. Thus v is a C^k-map in $V\backslash Y$. □

Notation In the remaining part of this section, U denotes the open unit disk in $\mathbb{C}$.

Definition 3.1.8 Let X be a Hausdorff topological space and A be a real algebra of continuous complex-valued functions on X. A continuous map $F: U \to X$ is called a harmonic map if $\mathrm{Re}\,(f \circ F)$ is harmonic on U for all f in A; F is called an analytic map (respectively, an antianalytic map) if the function $f \circ F$ is analytic (respectively, antianalytic) on U for all f in A.

Clearly, every analytic (antianalytic) map is harmonic. We now discuss a partial converse.

Theorem 3.1.9 Let X be a Hausdorff topological space, A be a real algebra of continuous complex-valued functions on X, and $F: U \to X$ be a harmonic map. Then F is an analytic map or an antianalytic map on each connected component of $U \backslash F^{-1}(Y)$, where

$$Y := \{x \in X : \operatorname{Im} g(x) = 0 \quad \text{for all } g \text{ in } A\}.$$

Proof. Let $\tilde{f} = f \circ F$ for f in A and $\tilde{A} = \{\tilde{f} : f \in A\}$. Then $\tilde{A}$ is closed under multiplication. Since $\operatorname{Re} \tilde{f} = \operatorname{Re}(f \circ F)$ is harmonic, it is a C^2-map on U for each $\tilde{f} \in \tilde{A}$. Hence by Lemma 3.1.7, $\operatorname{Im} \tilde{f}$ is a C^2-map on $U \backslash F^{-1}(Y)$ for all $\tilde{f}$ in $\tilde{A}$. Now let G be a connected component of $U \backslash F^{-1}(Y)$. Then it is easy to see that each $\tilde{f} \in \tilde{A}$ satisfies the hypotheses of Corollary 3.1.5. Hence $\tilde{f}$ is analytic or antianalytic. Therefore, by Lemma 3.1.6, all functions in $\tilde{A}$ are analytic on G, or all functions in $\tilde{A}$ are antianalytic on G. $\square$

Corollary 3.1.10
(i) Let X be a compact Hausdorff space, B a complex function algebra on X, and $F: U \to X$ a harmonic map. Then F is analytic or antianalytic on U.
(ii) Let τ be a topological involution on X, A a real function algebra on (X, τ), and $F: U \to X$ a harmonic map. Then F is analytic or antianalytic on each of the connected components of $U \backslash F^{-1}(\{x \in X : \tau(x) = x\})$.

Proof. Since B contains the constant function i, the set $\{x \in X : \operatorname{Im} g(x) = 0$ for all g in $B\}$ is empty. Further by Lemma 1.3.9,

$$\{x \in X : \operatorname{Im} g(x) = 0 \quad \text{for all } g \text{ in } A\} = \{x \in X : \tau(x) = x\}.$$

Now the conclusions follow from Theorem 3.1.9. $\square$

Example 3.1.11 Let X be the annular region in $\mathbb{C}$ defined by $X := \{z \in \mathbb{C} : r \leq |z| \leq 1/r\}$ for some r with $0 < r < 1$, and let τ be the involution given by $\tau(z) = -1/\bar{z}$ for all z in X. Let A be a real function algebra on (X, τ). Note that τ has no fixed points in X. By Corollary 3.1.10, if $F: U \to X$ is a harmonic map, then F must be an analytic map or an antianalytic map on U.

Remark 3.1.12 In the case of a complex function algebra, every harmonic map is an analytic map or an antianalytic map by Corollary 3.1.10, whereas we do not know whether this is true in the case of a real function algebra. We could only prove that a harmonic map is an analytic map or an antianalytic map in each connected component of $U \backslash F^{-1}(Y)$. However, we point out that under the hypotheses of Theorem 3.1.9, the set $F^{-1}(Y)$ has an empty interior if F is nonconstant. (Note that if F is constant, F is trivially an analytic map.) For if $F^{-1}(Y)$ contains a disk D, then for f in A and $\tilde{f} = f \circ F = u + iv$, and $v = 0$ throughout D. Now it is easy to see that the function $\tilde{f} = u + iv$ satisfies the hypotheses of Corollary 3.1.5. Hence u has a uniquely determined (single-valued) harmonic conjugate u^* on U and $\tilde{f} = u \pm iu^*$ on U; that is, $u^* = \pm v$ on U. But this implies that $u^* \equiv 0$ on D. Since u^* is harmonic, $u^* \equiv 0$ on U. Hence $\tilde{f} = u$. But then $u = \mathrm{Re}(f \circ F)$ and $u^2 = \mathrm{Re}(f^2 \circ F)$ are harmonic. Hence u is constant in U; that is, $f \circ F$ is constant, a contradiction.

Theorem   3.1.13   Let X be a subset of $\mathbb{C}$ that is compact, has connected interior, and which is symmetric with respect to the real axis. Let $\tau: X \to X$ be defined by $\tau(z) = \bar{z}$ for all z in X. Suppose that A is a real function algebra on (X, τ) such that $\mathrm{Re}\, f$ is harmonic on the interior X^0 of X for all f in A. Then f is analytic for all f in A or antianalytic for all f in A on X^0.

Proof.   Let $G = \{z \in X^0 : \mathrm{Im}\, z > 0\}$, and $f = u + iv$ be in A. By Corollary 3.1.10, $v = \mathrm{Im} \pm f$ is C^1 and since G is connected, Corollary 3.1.5 shows that f is analytic or antianalytic on G.

Since $f(\bar{z}) = \overline{f(z)}$ for all z in X, Schwarz's reflection principle implies that f is analytic or antianalytic in the whole of X^0. Now Lemma 3.1.6 implies that f is analytic for all f on A in X^0, or f is antianalytic for all f in A on X^0.   □

Most of the results in this section were published in Kulkarni (1983) and Arundhathi and Kulkarni (1986). Theorem 3.1.3 incorporates some improvements suggested by W. K. Hayman and R. R. Simha.

3.2   CHARACTERIZATIONS OF GLEASON PARTS

In this section we shall define a Gleason part of a real function algebra and give some characterizations of it. Some of these characterizations depend on measure-theoretic concepts, which we briefly review.

Notation   For a real or complex normed linear space A, the dual space of A, that is, the space of all (real-valued, if A is a real linear space, complex-valued, if A is a complex linear space) continuous linear functionals on A, will be denoted by A'.

Let X be a compact Hausdorff space. By a *measure on* X, we mean a complex regular Borel measure on X. The space of all such measures will be denoted by $BM(X)$. The terms *real measure* and *positive measure* will have the usual meanings. For a measure μ, the measures $\bar{\mu}$, Re μ, and Im μ are defined in an obvious manner; $|\mu|$ denotes the associated positive total variation measure. Then $\|\mu\|$, defined as $|\mu|(X)$, is a norm on $BM(X)$. A *probability measure* is a positive measure μ with $\mu(X) = 1$. For x in X, δ_x denotes the unit point mass at x, also known as the *Dirac delta measure at x*. The *support of a measure* μ is defined as the set of all x such that $|\mu|(U) > 0$ for every open neighborhood U of x. This set is denoted by Supp μ. Equivalently, it is the complement of the largest open set V such that $|\mu|(V) = 0$. Thus Supp μ is a closed set. We shall have several occasions to use the following well-known theorem, which gives a method of identifying $BM(X)$ with $C(X)'$.

Riesz Representation Theorem   For every continuous linear functional ϕ on $C(X)$, there exists a unique measure μ on X such that $\phi(f) = \int_X f \, d\mu$ for every f in $C(X)$ and $\|\phi\| = \|\mu\|$.

A proof of this result can be found in Rudin (1966). It is easy to see that if $\phi(f)$ is real for all real f, then μ is real and if $\phi(f) \geq 0$ for all $f \geq 0$, then μ is positive. A similar theorem is also true for functionals on $C_\mathbb{R}(X)$. In this case, the associated measures are real.

We now prove an analogous theorem for functionals on $C(X,\tau)$. We shall make use of the properties of the map σ given in Theorem 1.3.5.

Theorem   3.2.1   Let X be a compact Hausdorff space and τ a topological involution on X. Let σ be the algebra involution on $C(X)$ induced by τ, as in Theorem 1.3.5.

(i) Let ϕ be a continuous linear functional on $C(X)$, and let $\phi^*(f) := \overline{\phi}(\sigma(f))$. Then ϕ^* is also a continuous linear functional on $C(X)$.

(ii) For a measure μ on X, $\mu^* := \bar{\mu} \circ \tau$ (i.e., $\mu^*(E) := \bar{\mu}(\tau(E))$ for every Borel subset E of X) is also a measure on X.

(iii) The maps $\phi \mapsto \phi^*$ and $\mu \mapsto \mu^*$ are isometric linear involutions on $C(X)'$ and $BM(X)$, respectively.

(iv) If ϕ in $C(X)'$ and μ in $BM(X)$ are such that $\phi(f) = \int_X f \, d\mu$ for all f in $C(X)$ and $\|\phi\| = \|\mu\|$, then the same relationship holds between ϕ^* and μ^*; in particular, $\phi = \phi^*$ if and only if $\mu = \mu^*$.

(v) For ϕ in $C(X)'$ with $\phi = \phi^*$, the restriction of ϕ to $C(X,\tau)$ is a real-valued continuous (real-linear) functional on $C(X,\tau)$. In fact, this restriction map is a real-linear isometry of $\text{Sym}(C(X)') := \{\phi \in C(X)' : \phi = \phi^*\}$ onto the dual space of $C(X,\tau)$.

(vi) For every real-valued continuous (real-linear) functional ψ on $C(X,\tau)$, there exists a unique measure μ on X such that $\mu^* = \mu$, $\|\mu\| = \|\psi\|$, and $\psi(f) = \int_X f \, d\mu$ for every f in $C(X,\tau)$.

Proof. (i), (ii), and (iii) are obvious. To prove (iv), we note from (iii): $\|\phi^*\| = \|\phi\| = \|\mu\| = \|\mu^*\|$. Further,

$$\phi^*(f) = \overline{\phi}(\sigma(f)) = \overline{\int_X \sigma(f)\,d\mu} = \int_X \overline{\sigma(f)}\,d\overline{\mu}$$

$$= \int_X (f \circ \tau)\,d\overline{\mu} = \int_X f d(\overline{\mu} \circ \tau) = \int_X f d\mu^*$$

for every f in $C(X)$. Clearly, $\phi = \phi^*$ if and only if $\mu = \mu^*$.

(v) Let ϕ be in $C(X)'$ with $\phi = \phi^*$. Then for every f in $C(X,\tau)$, $\sigma(f) = f$ and hence

$$\phi(f) = \phi^*(f) = \overline{\phi}(\sigma(f)) = \overline{\phi}(f).$$

Thus $\phi(f)$ is real. Hence the restriction of ϕ to $C(X,\tau)$ is a continuous real-linear functional on $C(X,\tau)$. Now let ψ be a continuous real-linear functional on $C(X,\tau)$. Recall that every h in $C(X)$ can be expressed uniquely as $h = f + ig$ with f,g in $C(X,\tau)$ (Theorem 1.3.5). Define $\phi(h) := \psi(f) + i\psi(g)$. Clearly, $\phi \in C(X)'$. Further, since $\psi(f)$ and $\psi(g)$ are real,

$$\phi^*(h) = \overline{\phi}(\sigma(h)) = \overline{\phi}(f - ig) = \overline{\psi(f) - i\psi(g)}$$

$$= \psi(f) + i\psi(g)$$

$$= \phi(h).$$

Thus $\phi^* = \phi$. This shows that the restriction map is onto the dual of $C(X,\tau)$. It only remains to prove that it is an isometry. For this, recall from Theorem 1.3.5 that the map $P(f) = (f + \sigma(f))/2$, f in $C(X)$, defines a projection of $C(X)_{\mathbb{R}}$, that is, $C(X)$ regarded as a real-linear space onto $C(X,\tau)$. Also, $\|P(f)\| \le \|f\|$ for all f in $C(X)$. Now, let $\phi \in \mathrm{Sym}(C(X)')$. Then $\phi(\sigma(f)) = \overline{\phi}^*(f) = \phi(f)$ for all f in $C(X)$. Hence

$$\|\phi\| \ge \|\phi\|_{C(X,\tau)} := \sup\{|\phi(g)| : g \in C(X,\tau), \quad \|g\| \le 1\}$$

$$= \sup\{|\phi(P(f))| : f \in C(X), \quad \|P(f)\| \le 1\}.$$

Since $\|P(f)\| \le \|f\|$ for all f in $C(X)$, we have

$$\|\phi\| \ge \sup\left\{\frac{|\phi(f) + \phi(\sigma(f))|}{2} : f \in C(X), \quad \|f\| \le 1\right\}$$

$$= \sup\{|\mathrm{Re}\,\phi(f)| : f \in C(X), \quad \|f\| \le 1\}$$

$$= \sup\{|\phi(h)| : h \in C(X), \quad \|h\| \le 1\}$$

$$= \|\phi\|.$$

The last but one equality follows by considering $f = \exp(-i\theta)h$, if $\phi(h) = r\exp(i\theta)$, so that $\|f\| = \|h\|$ and Re $\phi(f) = r = |\phi(h)|$.

(vi) This follows from (v), the Riesz representation theorem, and (iv). $\square$

Statement (vi) of Theorem 3.2.1 can be considered as an analog of the Riesz representation theorem for $C(X,\tau)$. It was proved in Grzesiak (1989).

Definition 3.2.2 Let X be a compact Hausdorff space, A a real or complex subspace of $C(X)$, and $\phi \in A'$. A measure μ on X is called a *representing measure for ϕ* if $\|\mu\| = \|\phi\|$ and

$$\phi(f) = \int_X f\,d\mu \qquad \text{for all } f \text{ in } A.$$

Theorem 3.2.3
(a) Let A be a (real) subspace of $C_\mathbb{R}(X)$ or a complex subspace of $C(X)$ and $\phi \in A'$. Then there exists a representing measure for ϕ.
(b) Let τ be a topological involution on X, A be a (real) subspace of $C(X,\tau)$, and $\psi \in A'$. Then there exists a representing measure μ for ψ such that $\mu^* = \mu$.

Proof.
(a) Let $\bar\phi$ be a Hahn–Banach extension of ϕ to $C_\mathbb{R}(X)$ or $C(X)$, as the case may be, so that $\|\bar\phi\| = \|\phi\|$. By the Riesz representation theorem for $C_\mathbb{R}(X)$ and $C(X)$ quoted earlier, there is a measure μ on X such that $\|\bar\phi\| = \|\mu\|$, and for all f in A,

$$\phi(f) = \bar\phi(f) = \int_X f\,d\mu.$$

Hence μ is a representing measure for ϕ.

(b) By part (vi) of Theorem 3.2.1, we proceed exactly as in the proof of (a) above and obtain a representing measure μ for ψ, which, in addition, satisfies $\mu^* = \mu$. $\square$

We now consider a special situation. Let A be a real or complex subspace of $C(X)$ and $1 \in A$. We define

$$K(A) := \{\phi \in A' : \phi(1) = 1 = \|\phi\|\}.$$

A linear functional ϕ on A is called *positive* if $\phi(f) \geq 0$ for every $f \in A$ with $f \geq 0$.

Lemma 3.2.4 Let A be a real or complex subspace of $C(X)$ containing 1 and ϕ be a linear functional with $\phi(1) = 1$. Then $\|\phi\| = 1$ if and only if $\mathrm{Re}\,\phi(f) \geq 0$ for every f in A with $\mathrm{Re}\,f \geq 0$.

In particular, if $\phi \in K(A)$, then we have the following:
(i) If $f \in A$ and f is real-valued, then $\phi(f)$ is real.
(ii) ϕ is a positive functional on A.
(iii) If we let $\mathrm{Re}\,\phi(\mathrm{Re}\,f) = \mathrm{Re}\,\phi(f)$ for f in A, then $\mathrm{Re}\,\phi$ is a well-defined functional on $\mathrm{Re}\,A$ and $\mathrm{Re}\,\phi \in K(\mathrm{Re}\,A)$.

Proof. Suppose that $\|\phi\| = 1$, $f \in A$, $\mathrm{Re}\,f \geq 0$, and $\phi(f) = a + ib$. For each positive t, let $f_t := 1 - tf$. Then $f_t \in A$ and

$$|f_t|^2 = (1 - t\,\mathrm{Re}\,f)^2 + t^2(\mathrm{Im}\,f)^2 = 1 - 2t\,\mathrm{Re}\,f + t^2|f|^2$$
$$\leq 1 + t^2|f|^2,$$

so that

$$(1 - ta)^2 + t^2b^2 = |\phi(f_t)|^2 \leq \|f_t\|^2 \leq 1 + t^2\|f\|^2,$$

and hence $t(|\phi(f)|^2 - \|f\|^2) \leq 2a$. Letting $t \to 0$, we see that $0 \leq a$, that is, $\mathrm{Re}\,\phi(f) \geq 0$.

Conversely, assume that $\mathrm{Re}\,\phi(f) \geq 0$ for every f in A with $\mathrm{Re}\,f \geq 0$. Let $g \in A$ and $\|g\| \leq 1$. Then $\mathrm{Re}(1 \pm g) \geq 0$. Hence $1 \pm \mathrm{Re}\,\phi(g) \geq 0$; that is, $|\mathrm{Re}\,\phi(g)| \leq 1$. If A is a real subspace, $\phi(g)$ is real by definition. Hence $|\phi(g)| = |\mathrm{Re}\,\phi(g)| \leq 1$. If A is a complex subspace and $\phi(g) = r\exp(i\theta)$, consider $h = \exp(-i\theta)g$. Then $h \in A$, $\|h\| = \|g\| \leq 1$, and $\phi(h) = \exp(-i\theta)\phi(g) = r$. Hence by what we have just proved,

$$|\phi(g)| = r = |\mathrm{Re}\,\phi(h)| \leq 1.$$

This shows that $\|\phi\| \leq 1$. Since $\phi(1) = 1$, we have $\|\phi\| = 1$.

Now let $\phi \in K(A)$. Then $\mathrm{Re}\,\phi(f) \geq 0$ for every f in A with $\mathrm{Re}\,f \geq 0$.

(i) Let $f \in A$ be real-valued. If A is a real subspace, $\phi(f)$ is real by definition. If A is a complex subspace, then $\pm\mathrm{Im}\,\phi(f) = \mathrm{Re}\,\phi(\mp if) \geq 0$ because $\mathrm{Re}(\mp if) = 0$. Thus $\mathrm{Im}\,\phi(f) = 0$; that is, $\phi(f)$ is real. This proves (i).

(ii) Next, let $f \in A$ and $f \geq 0$. Then f is real and hence $\phi(f)$ is real by (i). Also, $\mathrm{Re}\,f = f \geq 0$, so that $\phi(f) = \mathrm{Re}\,\phi(f) \geq 0$. This shows that ϕ is a positive functional.

(iii) If f and g are in A and $\mathrm{Re}\,f = \mathrm{Re}\,g$, then $\mathrm{Re}(f - g) = 0 = \mathrm{Re}(g - f)$. Hence $\mathrm{Re}\,\phi(f - g) \geq 0$ and $\mathrm{Re}\,\phi(g - f) \geq 0$; that is, $\mathrm{Re}\,\phi(f) = \mathrm{Re}\,\phi(g)$. This shows that $\mathrm{Re}\,\phi(\mathrm{Re}\,f) = \mathrm{Re}\,\phi(f)$ is a well-defined functional on $\mathrm{Re}\,A$.

If $\mathrm{Re}\,f \in \mathrm{Re}\,A$ and $-1 \leq \mathrm{Re}\,f \leq 1$, then $1 \pm \mathrm{Re}\,f \geq 0$, so that

$1 \pm \text{Re } \phi(f) \geq 0$; that is, $-1 \leq \text{Re } \phi(f) \leq 1$. This implies that $\text{Re } \phi \in K(\text{Re } A)$. $\square$

Theorem   3.2.5
(a) Let A be a real or complex subspace of $C(X)$ containing 1. If $\phi \in K(A)$ has a representing measure μ, then μ is a probability measure.
(b) Let A be a (real) subspace of $C(X,\tau)$ containing 1 and $\psi \in K(A)$. Then there is a representing measure μ on X such that $\mu \circ \tau = \mu$, and every representing measure for ψ is a probability measure.

Proof.
(a) Consider

$$\tilde{\phi}(f) = \int_X f \, d\mu \qquad \text{for } f \text{ in } C(X).$$

Then $\tilde{\phi}(1) = \phi(1) = 1$ and $\|\tilde{\phi}\| = \|\mu\| = \|\phi\| = 1$. Hence by Lemma 3.2.4, $\tilde{\phi}$ is a positive functional and the associated measure μ is positive. As $\mu(X) = \phi(1) = 1$, we see that μ is a probability measure.
  (b) By Theorem 3.2.3(b), there is a representing measure μ for ψ such that $\mu^* = \overline{\mu \circ \tau} = \mu$. But since μ must be a probability measure by (a), it is real. Hence $\mu \circ \tau = \mu$. $\square$

In the task of partitioning the carrier space of a real function algebra A, we need certain properties of measures associated with a functional $\text{Re } \phi$: $A \to \mathbb{R}$ for $\phi \in \text{Car}(A)$. We shall derive these properties in a more general setup which will also be useful in Chapters 4 and 5.

Definition   3.2.6   Let X be a compact Hausdorff space, τ a topological involution on X, A a (real) subspace of $C(X,\tau)$ containing 1, and $\psi \in K(A)$. A *real part representing measure (RPR measure) for* ψ is a measure μ on X such that $\|\mu\| = 1$, $\mu \circ \tau = \mu$, and

$$\psi(f) = \int_X \text{Re } f \, d\mu$$

for all f in A.
  In case τ is the identity map, $C(X,\tau) = C_{\mathbb{R}}(X)$ and $f = \text{Re } f$ for every $f \in A$, so that a RPR measure for $\psi \in K(A)$ is nothing but a representing measure for ψ.

Theorem   3.2.7   Let A be a (real) subspace of $C(X,\tau)$ containing 1.
(a) Let $\psi \in K(A)$. Then there exists a RPR measure for ψ and every such measure is a probability measure.

(b) Let ψ_1 and ψ_2 be in $K(A)$. Suppose that there exists $k > 1$ such that

$$\frac{\psi_1(f)}{k} \leq \psi_2(f) \leq k\psi_1(f)$$

for all f in A with $\mathrm{Re}\, f \geq 0$. Then there exist RPR measures λ_1 for ψ_1 and λ_2 for ψ_2 with $\lambda_1/k \leq \lambda_2 \leq k\lambda_1$.

Proof.

(a) We give three proofs, based on the Riesz representation theorem for $C(X,\tau)$, $C(X)$, and $C_{\mathbb{R}}(X)$.

For the first proof, we claim that μ is a RPR measure for ψ if and only if μ is a representing measure for ψ and $\mu \circ \tau = \mu$. This can be seen as follows. Let μ be a measure on X such that $\mu \circ \tau = \mu$ and f be in $C(X,\tau)$. Then $\mathrm{Im}\, f = -\mathrm{Im}\, f \circ \tau$ and

$$\int_X \mathrm{Im}\, f \, d\mu = -\int_X \mathrm{Im}\, f \circ \tau \, d\mu = -\int_X \mathrm{Im}\, f \, d(\mu \circ \tau)$$

$$= -\int_X \mathrm{Im}\, f \, d\mu,$$

so that $\int_X \mathrm{Im}\, f \, d\mu = 0$. Hence for every f in $C(X,\tau)$ we have

$$\int_X f \, d\mu = \int_X \mathrm{Re}\, f \, d\mu + i \int_X \mathrm{Im}\, f \, d\mu = \int_X \mathrm{Re}\, f \, d\mu.$$

The claim above is now obvious, and the proof is complete upon appealing to Theorem 3.2.5(b).

The second proof proceeds as follows. Let $C(X)_{\mathbb{R}}$ denote $C(X)$ regarded as a real linear space. Let $\tilde{\psi}$ be a Hahn–Banach extension of ψ to $C(X)_{\mathbb{R}}$, so that $\|\tilde{\psi}\| = \|\psi\| = 1$ and $\tilde{\psi}(1) = \psi(1) = 1$. Since $\tilde{\psi} \in K(C(X)_{\mathbb{R}})$, we remark that if $h \in C(X)$ is real-valued, then $\mathrm{Re} \pm ih = 0$ and hence $\tilde{\psi}(ih) = 0$ by Lemma 3.2.4. Define $\phi\colon C(X) \to \mathbb{C}$ as follows:

$$\phi(f) := \tilde{\psi}(f) - i\tilde{\psi}(if), \qquad f \text{ in } C(X).$$

Then it is easy to see that $\|\phi\| = \|\tilde{\psi}\| = 1$, and if $h \in C(X)$ is real-valued, then $\phi(h) = \tilde{\psi}(h)$. In particular, $\phi(1) = \tilde{\psi}(1) = 1$, so that $\phi \in K(C(X))$. Again by Lemma 3.2.4(ii), ϕ is a positive functional. The Riesz representation theorem for $C(X)$ shows that there is a probability measure ν on X such that

$$\phi(f) = \int_X f \, d\nu$$

for all f in $C(X)$. Now, for f in $C(X)$, $\tilde{\psi}(i \operatorname{Im} f) = 0 = \tilde{\psi}(i \operatorname{Re} f)$ by an earlier remark and hence

$$
\begin{aligned}
\tilde{\psi}(f) &= \tilde{\psi}(\operatorname{Re} f + i \operatorname{Im} f) \\
&= \tilde{\psi}(\operatorname{Re} f) + \tilde{\mu}(i \operatorname{Im} f) \\
&= \tilde{\psi}(\operatorname{Re} f) = \phi(\operatorname{Re} f) = \int_X \operatorname{Re} f \, d\nu.
\end{aligned}
$$

Let $\mu = (\nu + \nu \circ \tau)/2$. Then $\mu \circ \tau = \mu$ and μ is a positive measure on X. Also, $\|\mu\| = |\mu|(X) = \mu(X) = (\nu(X) + \nu(\tau(X)))/2 = 1$. Further, for f in A, $\operatorname{Re} f \circ \tau = \operatorname{Re} f$, so that

$$
\int_X \operatorname{Re} f \, d(\nu \circ \tau) = \int_X \operatorname{Re} f \circ \tau \, d\nu = \int_X \operatorname{Re} f \, d\nu.
$$

Thus we see that for every f in A, we have

$$
\psi(f) = \tilde{\psi}(f) = \int_X \operatorname{Re} f \, d\nu = \int_X \operatorname{Re} f \, d\mu.
$$

This shows that μ is a RPR measure for ψ. Finally, (ii) of Lemma 3.2.4 shows that every RPR measure for ψ is a probability measure.

For the third proof, we note from (iii) of Lemma 3.2.4 that $\psi_0(\operatorname{Re} f) = \psi(f), f \in A$, is a well-defined functional on $\operatorname{Re} A$ and $\psi_0 \in K(\operatorname{Re} A)$. Let θ be a Hahn–Banach extension of ψ_0 to $C_{\mathbb{R}}(X)$, so that $\theta \in K(C_{\mathbb{R}}(X))$. By Lemma 3.2.4(ii), θ is a positive functional on $C_{\mathbb{R}}(X)$ and hence by the Riesz representation theorem for $C_{\mathbb{R}}(X)$, there is a probability measure ν on X such that for all h in $C_{\mathbb{R}}(X)$,

$$
\int_X h \, d\nu = \theta(h).
$$

Hence for all f in A,

$$
\psi(f) = \psi_0(\operatorname{Re} f) = \theta(\operatorname{Re} f) = \int_X \operatorname{Re} f \, d\nu.
$$

Now, as in the second proof, it follows that $\mu := (\nu + \nu \circ \tau)/2$ is a RPR measure for ψ and that every such measure is a probability measure.

(b) Let $\theta_1 = (k\psi_1 - \psi_2)/(k - 1)$ and $\theta_2 = (k\psi_2 - \psi_1)/(k - 1)$. Note that $k > 1$ by hypothesis. Clearly, $\theta_1 \in A'$ and $\theta_1(1) = 1$. By hypothesis, $\theta_1(f) \geq 0$ for f in A with $\operatorname{Re} f \geq 0$. By Lemma 3.2.4, $\|\theta_1\| = 1$. Thus $\theta_1 \in K(A)$. Similarly, $\theta_2 \in K(A)$. Hence by part (a) there exist RPR measures

μ_1 and μ_2 for θ_1 and θ_2, respectively. Then it is easy to see that $\lambda_1 := (k\mu_1 + \mu_2)/(k + 1)$ and $\lambda_2 := (\mu_1 + k\mu_2)/(k + 1)$ are RPR measures for $\psi_1 = (k\theta_1 + \theta_2)/(k + 1)$ and $\psi_2 = (\theta_1 + k\theta_2)/(k + 1)$, respectively. Further,

$$k\lambda_1 - \lambda_2 = (k - 1)\mu_2 \geq 0 \qquad \text{and} \qquad k\lambda_1 - \lambda_2 = (k - 1)\mu_1 \geq 0.$$

$\square$

Unlike in part (a) of Theorem 3.2.7, if ψ in A' satisfies only $\psi(1) = 1$, but $\|\psi\| \neq 1$, then there may not exist any measure on X which represents Re ψ. This is so even if ψ is a positive functional.

For example, let $X = [0,1]$, $\tau(x) = 1 - x$ for x in X, $A = C(X,\tau)$, and for f in A,

$$\psi(f) = \operatorname{Re} f(0) + \operatorname{Im} f(0).$$

Then $1 \in A$, $\psi \in A'$, $\psi(1) = 1$, ψ is a positive functional on A, but $\|\psi\| = \sqrt{2}$. Define $g(x) = i(1 - 2x)$ for x in X. Then $g \in A$ and $\psi(g) = 1$. Let μ be any measure on X. Since Re $g = 0$, we see that $\int_X \operatorname{Re} g \, d\mu = 0$. Thus $\psi(g) \neq \int_X \operatorname{Re} g \, d\mu$. Also, we note that there is no measure μ on $[0,1]$ satisfying $\mu \circ \tau = \mu$ which represents ψ. For if μ is a measure on X satisfying $\mu \circ \tau = \mu$, then for every f in A, $\int_X f \, d\mu = \int_X \operatorname{Re} f \, d\mu$, as in the first proof of Theorem 3.2.7, and hence $1 = \psi(g) \neq \int_X g \, d\mu = 0$. Of course, as per Theorem 3.2.3(b), ψ does have a representing measure μ on X satisfying $\mu^* = \overline{\mu \circ \tau} = \mu$, namely, $\mu = (\delta_0 + \delta_1)/2 + (\delta_0 - \delta_1)/2i$, where δ_0 and δ_1 are the unit point mass measures at 0 and 1, respectively. (It can be checked directly that $\|\mu\| = \sqrt{2}$.)

In case τ is the identity map, part (b) of Theorem 3.2.7 reduces to a well-known result of Bishop (Browder, 1969, p. 135) about representing measures for functionals on subspaces of $C_{\mathbb{R}}(X)$.

Now suppose that B is a complex function algebra on X and $\phi \in \operatorname{Car}(B)$. Then ϕ has a representing measure by Theorem 3.2.3(a). Since $\phi(1) = 1 = \|\phi\|$ (Remark 1.2.8), every representing measure for ϕ is a probability measure by Theorem 3.2.5(a).

Next, let τ be a topological involution on X, A a real function algebra on X, and $\phi \in \operatorname{Car}(A)$. Then

$$\|\phi\| := \sup\{|\phi(f)| : f \in A, \|f\| < 1\} = 1,$$

by Lemma 1.2.2. Hence $\|\operatorname{Re} \phi\| = 1 = \operatorname{Re} \phi(1)$, so that by Theorem 3.2.7(a), there exists a RPR measure for Re ϕ, and every such measure is a probability measure. Clearly, μ is a RPR measure for Re ϕ if and only if it is a RPR measure for Re $(\bar{\phi})$.

We shall use the following notation throughout: For ϕ and ψ in $\operatorname{Car}(A)$,

$$\|\phi - \psi\| := \sup\{|\phi(f) - \psi(f)| : f \in A, \|f\| < 1\}$$

and

$$\|(\phi - \psi)(\overline{\phi} - \psi)\| := \sup\{|(\phi(f) - \psi(f))(\overline{\phi}(f) - \psi(f))| : f \in A,$$
$$\|f\| < 1\}.$$

Since $\|\phi\| = \|\psi\| = 1$, we see that

$$\|\phi - \psi\| \le 2, \|\overline{\phi} - \psi\| \le 2, \text{ and } \|(\phi - \psi)(\overline{\phi} - \psi)\| \le 4.$$

The following lemma is adapted from a similar result about complex function algebras [Theorem 2.2, Chapter VI of Gamelin (1969)].

Lemma 3.2.8 Let X be a compact Hausdorff space and τ a topological involution on X. Let A be a real function algebra on (X, τ). Suppose that ϕ and ψ are in Car(A) such that $\|(\phi - \psi)(\overline{\phi} - \psi)\| = 4$. Then there are disjoint Borel subsets E_1 and E_2 of X such that $\tau(E_1) = E_1$, $\tau(E_2) = E_2$, every RPR measure for Re ϕ is supported on E_1, and every RPR measure for Re ψ is supported on E_2; in particular, every RPR measure for Re ϕ and every RPR measure for Re ψ are mutually singular.

Proof. There exist f_n in A such that $\|f_n\| \le 1$ for all n and

$$|(\phi - \psi)(\overline{\phi} - \psi)(f_n)| \longrightarrow 4 \qquad \text{as } n \longrightarrow \infty.$$

Passing to subsequences, if necessary, we may assume that $\phi(f_n) \to a$ and $\psi(f_n) \to b$ as $n \to \infty$, where $|a| \le 1$, $|b| \le 1$, and

$$|(a - b)(\overline{a} - b)| = 4.$$

It can be easily seen that we must have $|a - b| = 2$ and $|\overline{a} - b| = 2$. Hence either $a = 1$ and $b = -1$, or $a = -1$ and $b = 1$. We assume the former. Therefore, we can choose f_n in A with $\|f_n\| < 1$ such that

$$|1 - \phi(f_n)| < \frac{1}{n^2}$$

and

$$|1 - \psi(f_n)| < \frac{1}{n^2}$$

for each n. Now let

$$E_1 = \{x \in X : \text{Re } f_n(x) \longrightarrow 1\}$$

and

$$E_2 = \{x \in X : \text{Re } f_n(x) \longrightarrow -1\}.$$

Then E_1 and E_2 are disjoint Borel subsets of X, and clearly, $\tau(E_1) = E_1$ and $\tau(E_2) = E_2$.

Now let μ be a RPR measure for Re ϕ. Since $1 - \mathrm{Re}\, f_n \geq 0$ for all n, we can apply the monotone convergence theorem to deduce the following:

$$\int_X \left[\sum_{n=1}^{\infty} (1 - \mathrm{Re}\, f_n) \right] d\mu = \sum_{n=1}^{\infty} \int_X (1 - \mathrm{Re}\, f_n)\, d\mu$$

$$= \sum_{n=1}^{\infty} [1 - \mathrm{Re}\, \phi(f_n)]$$

$$\leq \sum_{n=1}^{\infty} |1 - \phi(f_n)|$$

$$\leq \sum_{n=1}^{\infty} \frac{1}{n^2} < \infty.$$

Thus $\sum_{n=1}^{\infty} (1 - \mathrm{Re}\, f_n)$ is finite a.e. (μ), so that $\mathrm{Re}\, f_n \to 1$ a.e. (μ). Hence μ is supported on E_1. Similarly, we can prove that every RPR measure for Re ψ is supported on E_2. □

Note that in the proof above, we did not use the fact that $\mu \circ \tau = \mu$.

We now take up the question of partitioning the carrier space of A to look for an analytic structure in it. In the complex case, the *Blaschke factors* play a prominent role in the search for an analytic structure. If B is a complex function algebra, $f \in B$, and a is a complex number with $|a| < 1$, then the function $(f - a)(1 - \bar{a}f)^{-1}$ is called a *Blaschke factor*. The Blaschke factors play an important role in the real case also. But since A is only a real algebra, one has to consider a *joint Blaschke factor* as follows. For f in A with $\|f\| \leq 1$ and a in $\mathbb{C}$ with $|a| < 1$, let

$$\beta(f,a) := \frac{(f - a)(f - \bar{a})}{(1 - \bar{a}f)(1 - af)} = \frac{f^2 - (a + \bar{a})f + |a|^2}{|a|^2 f^2 - (a + \bar{a})f + 1}.$$

Note that $\beta(f,a) \in A$ and $\|\beta(f,a)\| \leq 1$.

Theorem 3.2.9 Let X be a compact Hausdorff space and τ a topological involution on X. Let A be a real function algebra on (X,τ), and ϕ, ψ be in Car(A). Then the following statements are equivalent:

(i) $\|(\phi - \psi)(\bar{\phi} - \psi)\| < 4$.

(ii) $\|\phi - \psi\| < 2$ or $\|\bar{\phi} - \psi\| < 2$.

(iii) $\sup\{|\psi(f)| : f \in A, \|f\| < 1, \phi(f) = 0\} < 1$.

(iv) If (f_n) is a sequence in A such that $\|f_n\| < 1$ for all n and $\phi(f_n) \to a$

as $n \to \infty$ with $|a| = 1$, then every convergent subsequence of $(\psi(f_n))$ converges to a or to $\bar{a}$.

(v) If (f_n) is a sequence in A such that $\|f_n\| < 1$ for all n and $|\phi(f_n)| \to 1$ as $n \to \infty$, then $|\psi(f_n)| \to 1$ as $n \to \infty$.

(vi) There is $k > 1$ such that for all f in A with $\mathrm{Re}\, f > 0$, we have $k^{-1} \mathrm{Re}\, \phi(f) \le \mathrm{Re}\, \psi(f) \le k\, \mathrm{Re}\, \phi(f)$.

(vii) There are RPR measures μ and λ on X for $\mathrm{Re}\, \phi$ and $\mathrm{Re}\, \psi$, respectively, such that $k^{-1}\mu \le \lambda \le k\mu$ for some $k > 0$.

Proof.

(i) implies (ii): Let (i) hold. If (ii) does not hold, then since $\|\phi - \psi\| = 2$ and $\|\phi\| = 1 = \|\psi\|$, there is a sequence (f_n) in A such that $\|f_n\| < 1$ for all n, $\phi(f_n) \to a$ and $\psi(f_n) \to -a$ as $n \to \infty$, where $|a| = 1$.

First assume that $a \ne \pm 1, \pm i$. For $n = 1,2,\dots$ and $0 < r < 1$, let $f_{n,r} = \beta(f_n, ra)$. Then $f_{n,r} \in A$ and $\|f_{n,r}\| < 1$ for all n and r. Also, as $n \to \infty$,

$$\phi(f_{n,r}) \longrightarrow \frac{a^2 - r(a + \bar{a}) + r^2|a|^2}{1 - r(a + \bar{a}) + r^2|a|^2 a^2} = \frac{a^2 - r}{1 - ra^2},$$

which, in turn, tends to -1 as $r \to 1$. (Note that $a \ne \pm 1$.) On the other hand, as $n \to \infty$,

$$\psi(f_{n,r}) \longrightarrow \frac{a^2 + r(a + \bar{a}) + r^2|a|^2}{1 + r(a + \bar{a}) + r^2|a|^2 a^2} = \frac{a^2 + r}{1 + ra^2},$$

which, in turn, tends to 1 as $r \to 1$. (Note that $a \ne \pm i$.) It now follows that

$$|(\phi - \psi)(\bar{\phi} - \psi)(f_{n,r})| \longrightarrow 4$$

as $n \to \infty$ and $r \to 1$, which is a contradiction to (i).

Next, let $a = \pm 1$. Then

$$|(\phi - \psi)(\bar{\phi} - \psi)(f_n)| \longrightarrow 4$$

as $n \to \infty$, which is also a contradiction to (i).

Finally, let $a = \pm i$. Since $\|\bar{\phi} - \psi\| = 2$, we may find g_n in A such that $\|g_n\| < 1$ for all n, $\bar{\phi}(g_n) \to b$ and $\psi(g_n) \to -b$ as $n \to \infty$, where $|b| = 1$. The cases $b \ne \pm 1, \pm i$ and $b = \pm 1$ can be treated as above. If $b = \pm i$, then let $h_n = f_n g_n$. Since $a = \pm i$, we can easily verify that

$$|(\phi - \psi)(\bar{\phi} - \psi)(h_n)| \longrightarrow 4$$

as $n \to 1$. Note that $h_n \in A$ and $\|h_n\| < 1$. This again contradicts (i).

(ii) implies (iii): Suppose that $\|\phi - \psi\| = 2c$, where $0 \le c < 1$. If (iii) does not hold, there exists a sequence (f_n) in A such that $\|f_n\| < 1$,

$\phi(f_n) = 0$ for all n, and $\psi(f_n) \to a$ as $n \to \infty$, where $|a| = 1$.

First, suppose that $a \neq \pm 1$. Define for $n = 1, 2, \ldots$ and $0 < r < 1$, $f_{n,r} = \beta(f_n, ra)$. Then $f_{n,r} \in A$ and $\|f_{n,r}\| < 1$ for all n and r. Also, $\phi(f_{n,r}) = r^2 \to 1$ as $r \to 1$, and as we have earlier seen, $\psi(f_{n,r}) \to -1$ as $n \to \infty$ and $r \to 1$. This contradicts (ii).

Now, let $a = \pm 1$. Considering $(-f_n)$ in place of (f_n), if necessary, we can assume that $a = 1$. For $n = 1, 2, \ldots$, define

$$g_n := \frac{c - f_n}{1 - cf_n}.$$

Then $g_n \in A$ and $\|g_n\| < 1$ for all n. Hence

$$|\phi(g_n) - \psi(g_n)| \leq 2c,$$

that is,

$$\left| c - \frac{c - \psi(f_n)}{1 - c\psi(f_n)} \right| \leq 2c$$

for all n. Letting $n \to \infty$, we see that $1 + c \leq 2c$, which is a contradiction to our assumption $c < 1$. Similar proof holds if $\|\bar{\phi} - \psi\| = 2c$ with $0 \leq c < 1$.

(iii) implies (iv): Let (f_n) be a sequence in A such that $\|f_n\| < 1$ for all n and $\phi(f_n) \to a$ as $n \to \infty$, where $|a| = 1$. If a subsequence of $(\psi(f_n))$ [which we shall denote by $(\psi(f_n)$ itself] converges to b, where b is different from a and $\bar{a}$, then let $g_n := \beta(f_n, \phi(f_n))$. Now $g_n \in A$, $\|g_n\| < 1$, and $\phi(g_n) = 0$ for all n, while

$$\psi(g_n) \longrightarrow \frac{b^2 - (a + \bar{a})b + |a|^2}{|a|^2 b^2 - (a + \bar{a})b + 1} = 1$$

as $n \to \infty$. This contradicts (iii).

(iv) implies (v): The proof follows by passing to subsequences sufficiently many times.

(v) implies (vi): If (vi) does not hold, then there is a sequence (p_n) in A with Re $p_n > 0$ for all n such that either

$$\frac{\text{Re } \psi(p_n)}{\text{Re } \phi(p_n)} \longrightarrow \infty \quad \text{or} \quad \frac{\text{Re } \phi(p_n)}{\text{Re } \psi(p_n)} \longrightarrow \infty$$

as $n \to \infty$. In the first case, we may find a sequence (α_n) of positive real numbers such that $\alpha_n \to 0$ but

$$\alpha_n \frac{\text{Re } \psi(p_n)}{\text{Re } \phi(p_n)} \longrightarrow \infty$$

and let

$$f_n := \frac{\alpha_n p_n}{\operatorname{Re} \phi(p_n)}.$$

In the second case, we may find a sequence (β_n) of positive real numbers such that $\beta_n \to 0$ but

$$\beta_n \frac{\operatorname{Re} \phi(p_n)}{\operatorname{Re} \psi(p_n)} \longrightarrow \infty$$

and let

$$f_n := \frac{\beta_n p_n}{\operatorname{Re} \psi(p_n)}.$$

Thus there is a sequence (f_n) in A with $\operatorname{Re} f_n > 0$ for all n such that either (1) $\operatorname{Re} \phi(f_n) \to 0$ and $\operatorname{Re} \psi(f_n) \to \infty$ as $n \to \infty$, or (2) $\operatorname{Re} \phi(f_n) \to \infty$ and $\operatorname{Re} \psi(f_n) \to 0$ as $n \to \infty$. Define $g_n := \exp(-f_n)$. Then $g_n \in A$, $\|g_n\| < 1$,

$$|\phi(g_n)| = \exp(-\operatorname{Re} \phi(f_n)) \quad \text{and} \quad |\psi(g_n)| = \exp(-\operatorname{Re} \psi(f_n)).$$

In case alternative (1) holds, we have $|\phi(g_n)| \to 1$ with $|\psi(g_n)| \to 0$ as $n \to \infty$, in contradiction to (v). In case alternative (2) holds, let $h_n := b(g_n, \psi(g_n))$. Then $\psi(h_n) = 0$ and $|\phi(h_n)| \to 1$ as $n \to \infty$, again in contradiction to (v).

(vi) implies (vii): This follows from Theorem 3.2.7(b) since the inequalities in (vi) are easily seen to hold for all f in A with $\operatorname{Re} f \geq 0$.

(vii) implies (i): This follows from Lemma 3.2.8. □

Definition 3.2.10 Let X be a compact Hausdorff space and τ be a topological involution on X.

For a real function algebra A on (X, τ) and ϕ, ψ in $\operatorname{Car}(A)$, define $\phi \sim \psi$ if any one of the equivalent statements (i) to (vii) of Theorem 3.2.9 holds. From statement (vi) it can be seen that $\sim$ is an equivalence relation on $\operatorname{Car}(A)$. We call an equivalence class under $\sim$, a *Gleason part of A*. We shall denote the equivalence class containing ϕ by $Q_A(\phi)$. Obviously, $\overline{\psi} \in Q_A(\phi)$ for each ψ in $Q_A(\phi)$.

Example 3.2.11 Let X be a compact Hausdorff space and τ be a topological involution on X. Let $A = C(X, \tau)$. Then $\operatorname{Car}(A) = \{e_x : x \in X\}$ (Corollary 1.3.21). Now if $x, y \in X$ and $y \notin \{x, \tau(x)\}$, then by Urysohn's lemma, we can find f in A such that $0 \leq f \leq 1$, $f(x) = 0 = f(\tau(x))$ and $f(y) = 1$.

Then, by Theorem 3.2.9(iii), $e_y \notin Q_A(e_x)$. Thus Gleason parts of A are the sets $\{x, \tau(x)\}$ for x in X.

Gleason parts are often referred to as simply *parts*. A part consisting of more than two points is called *nontrivial*.

We now consider the parts of the real disk algebra A of Example 1.3.10. Let $|t| < 1, f \in A$, and $\|f\| < 1$. Let z be the function considered in Example 1.3.10: $z(s) = s$ for all s. Then $(f - f(t))/(z - t)$ has a removable singularity at t. Hence the function g defined by

$$g := \frac{[f - f(t)](1 - \bar{t}z)}{(z - t)[1 - \overline{f(t)}f]}$$

is analytic on the open and continuous on the closed unit disk. By the maximum modulus principle, for every s in the closed unit disk X, we have

$$|g(s)| \le \max\{|g(w)| : |w| = 1\} \le 1.$$

Hence

$$|f(s) - f(t)| \le \left|\frac{s - t}{1 - \bar{t}s}\right| |1 - \overline{f(t)}f(s)|.$$

This is essentially a reformulation of the classical Schwarz lemma. Now if $|s| < 1$, then the first factor on the right-hand side is strictly less than 1. Hence

$$\sup\{|f(s) - f(t)| : f \in A, \|f\| < 1\} < 2.$$

Thus e_s and e_t belong to the same part. On the other hand, this supremum is 2 if $|s| = 1, |t| \le 1$, and $t \ne s, t \ne \bar{s}$: Let f_r denote the joint Blaschke factor $\beta(z, rs)$ for $0 < r < 1$. Then $f_r \in A$, $\|f_r\| = 1$, and as $r \to 1$,

$$f_r(s) = \frac{s^2 - r}{1 - rs^2} \longrightarrow -1$$

and

$$f_r(t) \longrightarrow \frac{t^2 - (s + \bar{s})t + |s|^2}{1 - (s + \bar{s})t + |s|^2t^2} = 1,$$

since $t \ne s, t \ne \bar{s}$. Thus, identifying x with e_x for each x in X, we can say that the open unit disk is a nontrivial part, $\{s, \bar{s}\}$ are two-point parts for each s satisfying $|s| = 1$ and $s \ne \pm 1$, while $\{1\}$ and $\{-1\}$ are one-point parts.

Consider a real function algebra A on a compact Hausdorff space X

with a topological involution τ. If there exists an analytic map (Definition 3.1.8) F of the open unit disk U into Car(A), then $F(U)$ is called an *analytic disk* in Car(A). Now, if $\phi = F(s)$, $\psi = F(t)$ for some s, t in U, and if $f \in A$ with $\|f\| < 1$, then $\hat{f} = f \circ F$ is an analytic function on U, and by Schwarz's lemma,

$$|\phi(f) - \psi(f)| = |\hat{f}(s) - \hat{f}(t)| \le \left|\frac{s-t}{1 - \bar{t}s}\right| |1 - \hat{f}(s)\bar{\hat{f}}(t)|$$

$$\le 2\left|\frac{s-t}{1 - \bar{t}s}\right|.$$

Hence

$$\|\phi - \psi\| = \sup\{|\phi(f) - \psi(f)| : f \in A, \|f\| < 1\} < 2.$$

Thus an analytic disk must be contained in a single Gleason part of A. More generally, if any two points of Car(A) can be connected by a finite chain of analytic disks, then by the transitivity of $\sim$, these points belong to the same part. In Examples 1.3.11 and 1.3.12, the functions in A are analytic in the interior of X. Hence, identifying x with e_x, we can say that every point in the interior of X is contained in an analytic disk. By the argument above, for each connected component G of the interior of X, $G \cup \tau(G)$ is contained in a nontrivial Gleason part. In particular, in Example 1.3.12, $X = \{z \in \mathbb{C} : c \le |z| \le 1/c\}$ for some c with $0 < c < 1$ and $\tau(z) = -1/\bar{z}$. Thus $X^0 = \{z \in \mathbb{C} : c < |z| < 1/c\}$ is a nontrivial part.

Similar comments can be made about antianalytic maps and *antiana-lytic disks*. Now let F be a harmonic map of U into Car(A). We call $F(U)$ a *harmonic disk* in Car(A). Let $Y = \{\phi \in \text{Car}(A) : \phi = \bar{\phi}\}$. Then by Corollary 3.1.10, F is an analytic or antianalytic map on each of the connected components of $U \setminus F^{-1}(Y)$. Hence each connected component of $F(U) \setminus Y$ must lie in a single Gleason part.

Corollary 3.2.12   Let A be a real function algebra on (X, τ) and K a weak peak set for A (Definition 2.2.1). If $x \in K$ and $y \notin K$, then x and y (i.e., e_x and e_y) belong to different Gleason parts of A. In particular, if $K = \{x, \tau(x)\}$, then it constitutes a (trivial) part of A.

Proof.   Since K is compact (Remark 2.2.2), one can find an open neighborhood U of K such that U does not contain y. Let $0 < \varepsilon < 1$. By Theorem 2.2.3(iii), there exists f in A such that $\|f\| = 1$, $f_{|K} \equiv 1$, and $\|f\|_{X|U} < \varepsilon$. Thus $f(x) = 1$ and $|f(y)| < \varepsilon$. Now consider $f_n := (1 - 1/n)f$ for all n. Then $\|f_n\| < 1$ and $|f_n(y)| < \varepsilon$ for all n, while $f_n(x) \to 1$ as $n \to \infty$. Hence by Theorem 3.2.9(iv), x and y belong to different parts. $\square$

Remark 3.2.13 Let Q be a Gleason part of a real function algebra A on (X,τ). Then the map $\phi \mapsto \bar{\phi}$ is a topological involution on Q. Since Q is a subset of the compact Hausdorff space $\mathrm{Car}(A)$, Q is completely regular. We now show that Q is σ-compact. Let $\phi \in Q$. For natural numbers n and m, we define

$$A_n := \left\{ \psi \in \mathrm{Car}(A) : \|\phi - \psi\| \leq 2 - \frac{1}{n} \right\}$$

and

$$B_m := \left\{ \psi \in \mathrm{Car}(A) : \|\bar{\phi} - \psi\| \leq 2 - \frac{1}{m} \right\}.$$

Then A_n and B_m are compact. By Theorem 3.2.9(ii), $Q = (\bigcup_n A_n) \cup (\bigcup_m B_m)$. Thus Q is σ-compact. We shall show in Section 3.5 that these topological properties characterize a Gleason part.

Let A be a real commutative Banach algebra with unit. We have observed in Remark 1.2.7 that the map $\ker : \mathrm{Car}(A) \to M(A)$ is onto. Hence every equivalence relation on $\mathrm{Car}(A)$ can be carried over to $M(A)$. This is formalized in the next corollary. Recall the definitions of the maps $\mathrm{Re}\,\hat{f}$, $|\hat{f}|$ on $M(A)$ for f in A, given in Remark 1.2.11.

Corollary 3.2.14 Let A be a real uniform algebra (Definition 1.3.13), and $x, y \in M(A)$. Then the following are equivalent:
(i) $\sup\{|\hat{f}|(y) : f \in A, \|f\| < 1, |\hat{f}|(x) = 0\} < 1$.
(ii) If (f_n) is a sequence in A such that $\|f_n\| < 1$ for all n and $|\hat{f_n}|(x) \to 1$ as $n \to \infty$, then $|\hat{f_n}|(y) \to 1$ as $n \to \infty$.
(iii) There is a constant $k > 1$ such that for all $f \in A$ with $\mathrm{Re}\,\hat{f} > 0$, we have

$$k^{-1} \mathrm{Re}\,\hat{f}(x) \leq \mathrm{Re}\,\hat{f}(y) \leq k\,\mathrm{Re}\,\hat{f}(x).$$

(iv) There exist regular Borel probability measures $\hat{\mu}$ and $\hat{\nu}$ on $M(A)$ and a constant $k > 1$ such that

$$\int_{M(A)} \mathrm{Re}\,f\,d\hat{\mu} = \mathrm{Re}\,\hat{f}(x) \quad \text{and} \quad \int_{M(A)} \mathrm{Re}\,f\,d\hat{\nu} = \mathrm{Re}\,\hat{f}(y)$$

for all f in A and $k^{-1}\hat{\mu} \leq \hat{\nu} \leq k\hat{\mu}$.

Proof. By Theorem 1.3.7, A can be regarded as a real function algebra on $(\mathrm{Car}\,A, \tau)$. Let ϕ, ψ in $\mathrm{Car}(A)$ be such that $x = \ker\phi$ and $y = \ker\psi$. Then for every f in A, $\mathrm{Re}\,\hat{f}(x) = \mathrm{Re}\,\phi(f)$, $\mathrm{Re}\,\hat{f}(y) = \mathrm{Re}\,\psi(f)$, $|\hat{f}|(x) = |\phi(f)|$, and $|\hat{f}|(y) = |\phi(f)|$. Hence the equivalence of statements (i), (ii), and (iii) follows from the equivalence of statements (iii), (v), and (vi) of Theorem 3.2.9. Next, (iv) obviously implies (iii), while (iii) implies (iv) by Theorem 3.2.7(b) by letting $X = M(A)$, τ the identity map on X, and considering the subspace $\{\mathrm{Re}\,\hat{f} : f \in A\}$ of $C_{\mathbb{R}}(X)$. $\qquad\square$

Definition 3.2.15 For a real uniform algebra A and x, y in $M(A)$, we see that statement (iii) of Corollary 3.2.14 defines an equivalence relation on $M(A)$. The equivalence class of x in $M(A)$ under this relation is denoted by $P_A(x)$. It is clear that $\ker(Q_A(\phi)) = P_A(\ker \phi)$ for every ϕ in $\mathrm{Car}(A)$.

Corollary 3.2.16 Let A be a complex function algebra on a compact Hausdorff space X and $\phi, \psi \in \mathrm{Car}(A)$. Then the following statements are equivalent:
(i) $\|\phi - \psi\| < 2$.
(ii) $\sup\{|\psi(f)| : f \in A, \|f\| < 1, \phi(f) = 0\} < 1$
(iii) If $\{f_n\}$ is a sequence in A such that $\|f_n\| < 1$ for all n and $\phi(f_n) \to a$ as $n \to \infty$ with $|a| = 1$, then every convergent subsequence of $\{\psi(f_n)\}$ converges to a.
(iv) If $\{f_n\}$ is a sequence in A such that $\|f_n\| < 1$ for all n and $|\phi(f_n)| \to 1$ as $n \to \infty$, then $|\psi(f_n)| \to 1$ as $n \to \infty$.
(v) There is $k > 1$ such that for all f in A with $\mathrm{Re}\,f > 0$, we have $k^{-1}\,\mathrm{Re}\,\phi(f) \le \mathrm{Re}\,\psi(f) \le k\,\mathrm{Re}\,\phi(f)$.
(vi) There are representing measures μ and λ on X for ϕ and ψ, respectively, such that $k^{-1}\mu \le \lambda \le k\mu$ for some $k > 1$.

Proof. By Remark 1.3.19, A can be regarded as a real function algebra A_R on (Y, τ), where Y is a disjoint union of two copies of X and τ sends a point in one copy to the corresponding point in the other copy. Then $\mathrm{Car}(A) \subset \mathrm{Car}(A_R)$ by Remark 1.2.8. Now by Theorem 3.2.9, statements (i), (ii), (iv), and (v) are equivalent. Also, (iii) implies (iv), and (vi) implies (v) obviously. Further, (v) implies (vi) by Theorem 3.2.7(b). We now prove (ii) implies (iii). Let $\{f_n\}$ be as in (iii). Then $\phi(\bar{a}f_n) \to 1$ as $n \to \infty$. Hence by Theorem 3.2.9(iv), $\psi(\bar{a}f_n) \to 1$; that is, $\psi(f_n) \to a$ as $n \to \infty$. $\qquad\square$

Remark 3.2.17 We see that statement (v) in Corollary 3.2.16 defines an equivalence relation on the carrier space of a complex function algebra A. Using Remark 1.2.8, we shall identify the carrier space of A with the maximal ideal space $M(A)$ of A, and denote the equivalence class containing ϕ by $P_A(\phi)$ for ϕ in $\mathrm{Car}(A)$. These equivalence classes are called *Gleason parts* (or simply *parts*) of the complex function algebra A. A part containing more than one point is called *nontrivial*. The foregoing characterizations of Gleason parts of a complex function algebra are well known [see, e.g., Browder (1969) and Gamelin (1969)]. Theory of Gleason parts of real function algebras was developed in Kulkarni and Limaye (1981b).

As in Example 3.2.11, a simple application of Urysohn's lemma shows that all the parts of $C(X)$ are trivial. If B is the (complex) disk algebra of Example 1.3.10, then we can show that $\|e_s - e_t\| < 2$ if $|s| < 1, |t| < 1$. If $|s| = 1, |t| \leq 1$, and $t \neq s$, we can consider the functions $h_r = \bar{s}(z - rs)/(1 - r\bar{s}z)$ to show that $\|e_s - e_t\| = 2$. Thus the open unit disk is a nontrivial part of B and $\{s\}$ is a trivial part for each s with $|s| = 1$. Now, following the argument after Example 3.2.11, one can deduce that for algebras B of Examples 1.3.11 and 1.3.12, each connected component of the interior of X is a nontrivial part and each point of the boundary of X is a trivial part.

Corollary 3.2.18 Let A be a real or complex function algebra, $\phi \in$ Car(A), and $f \in A$ such that $\|f\| = |\phi(f)|$. Then f is a constant on the Gleason part Q containing ϕ if A is a complex algebra, and f takes at most two values on Q if A is a real algebra.

Proof. We may assume that $\|f\| = 1$ and $\phi(f) = a$ with $|a| = 1$. Let $f_n := (1 - 1/n)f, n = 1,2,\ldots$. Then for each ψ in Q,

$$\hat{f}(\psi) := \psi(f) = \lim_{n \to \infty} \psi(f_n) = \begin{cases} a & \text{if } A \text{ is a complex} \\ a \text{ or } \bar{a} & \text{if } A \text{ is a real.} \end{cases}$$

The last equality follows from Corollary 3.2.16(iii) and Theorem 3.2.9(iv). $\qquad \square$

Remark 3.2.19 For a complex function algebra A and ϕ, ψ in Car(A), let

$g(\phi,\psi) := \|\phi - \psi\| = \sup\{|\phi(f) - \psi(f)| : f \in A, \quad \|f\| \leq 1\}$.

$d(\phi,\psi) := \sup\{|\phi(f)| : f \in A, \quad \|f\| < 1, \quad \psi(f) = 0\}$,

$h(\phi,\psi) := \ln(\inf\{k \geq 1 : k^{-1} \operatorname{Re} \psi(f) \leq \operatorname{Re} \phi(f) \leq k \operatorname{Re} \psi(f)$

$$\text{for all } f \in A \text{ with } \operatorname{Re} f > 0\}.)$$

Corollary 3.2.16 shows that $g(\phi,\psi) < 2$ if and only if $d(\phi,\psi) < 1$ if and only if $h(\phi,\psi) < \infty$, and in that case ϕ and ψ belong to the same Gleason part. König (1969) has shown that d and h are metrics on a Gleason part of A, and each is equivalent to the metric g; in fact,

$$h(\phi,\psi) = \ln\left[\frac{1 + d(\phi,\psi)}{1 - d(\phi,\psi)}\right] = 2\ln\left[\frac{2 + g(\phi,\psi)}{2 - g(\phi,\psi)}\right]$$

[see Bear (1965, 1970) as well].

Now, let A be a real function algebra. For $\phi, \psi \in \text{Car}(A)$, definitions of g, d, and h above are meaningful. Theorem 3.2.9 shows that $g(\phi, \psi) < 2$ or $g(\overline{\phi}, \psi) < 2$ if and only if $d(\phi, \psi) < 1$ if and only if $h(\phi, \psi) < \infty$, and in that case ϕ and ψ belong to the same part. While g defines a metric on a Gleason part of A, h defines only a pseudometric on it since $h(\phi, \overline{\phi}) = 0$, and d may not even be symmetric. Let A be the real disk algebra and identify $\text{Car}(A)$ with the closed unit disk in $\mathbb{C}$. Then $d(i/2, 0) = \frac{1}{2}$ by Schwarz's lemma, but $d(0, i/2) \leq \frac{1}{4}$, since $f \in A$, $\|f\| < 1$, $f(i/2) = 0$ imply that $f(-i/2) = 0$ and the function g given by

$$g(z) := \frac{f(z)(1 + iz/2)(1 - iz/2)}{(z - i/2)(z + i/2)}, \qquad |z| \leq 1$$

is continuous on the closed unit disk and analytic on the open unit disk, so that $4|f(0)| = |g(0)| < 1$.

Notice that if A is a complex function algebra and ϕ, ψ are in $\text{Car}(A)$, then

$$d(\phi, \psi) = \sup \left\{ \frac{|\phi(f) - \psi(f)|}{|1 - \overline{\psi(f)}\phi(f)|} : f \in A, \ \|f\| < 1 \right\}.$$

This follows by considering the function $(f - \psi(f))/(1 - \overline{\psi(f)}f)$ for f in A with $\|f\| < 1$. This procedure may not work for a real function algebra unless $\overline{\psi} = \psi$. Hence let

$$d_0(\phi, \psi) = \sup \left\{ \frac{|\phi(f) - \psi(f)| \, |\phi(f) - \overline{\psi(f)}|}{|1 - \overline{\psi(f)}\phi(f)| \, |1 - \psi(f)\phi(f|} : f \in A, \ \|f\| < 1 \right\}.$$

Then it is easily seen that $d_0(\phi, \psi) = d_0(\psi, \phi)$ and $d(\phi, \psi) \leq \sqrt{d_0(\phi, \psi)}$. On the other hand, if $f \in A$ and $\|f\| < 1$, then by considering the joint Blaschke factor $\beta(f, \psi(f))$, it follows that $d_0(\phi, \psi) \leq d(\phi, \psi)$. In particular, if $d(\phi, \psi) = 0$, then for a given f in A, either $\phi(f) = \psi(f)$ or $\phi(f) = \overline{\psi(f)}$, so that either $\phi = \psi$ or $\phi = \overline{\psi}$.

Also, for ϕ, ψ in $\text{Car}(A)$,

$$d(\phi, \psi) \leq \sqrt{d_0(\phi, \psi)} = \sqrt{d_0(\psi, \phi)} \leq \sqrt{d(\psi, \phi)}.$$

These results are proved by Kulkarni and Arundhathi (1991b). They also prove the following result:

Let A be a real function algebra, and ϕ and ψ belong to the same Gleason part of A.

(a) Let $r > 0$ and $k(r) = (\exp(r) - 1)/\exp(r + 1)$.

If $h(\phi,\psi) \geq r$, then

$$\min\{g(\phi,\psi), g(\phi,\overline{\psi})\} > k(r) \qquad \text{and} \qquad d(\phi,\psi) > \frac{k(r)^2}{4}.$$

(b) $d(\phi,\overline{\psi}) \geq 16g(\phi,\psi)g(\phi,\overline{\psi})/[4 + g(\phi,\psi)^2][4 + g(\phi,\overline{\psi})^2]$.

3.3 PARTS OF THE COMPLEXIFICATION

Let X be a compact Hausdorff space and τ be a topological involution on X. Let A be a real function algebra on (X,τ) and $B = \{f + ig : f,g \in A\}$. We have shown in Theorem 1.3.20 that B is a complex function algebra on X and can be regarded as a complexification of A. In this section we study the relationship between the Gleason parts of A and those of B. The following facts proved in Theorem 1.3.20 will be used repeatedly in the derivation of our main results. For f,g in A, $\|f + ig\| = \|f - ig\|$,

$$\max\{\|f\|, \|g\|\} \leq \|f + ig\| \leq \|f\| + \|g\|.$$

The map $\alpha \colon \mathrm{Car}(A) \to \mathrm{Car}(B)$ defined by

$$\alpha(\phi)(f + ig) = \phi(f) + i\phi(g)$$

for ϕ in $\mathrm{Car}(A)$ and f,g in A is a bijection and $\alpha(\phi)|_A = \phi$.

Also, since for ϕ,ψ in $\mathrm{Car}(A)$ and f,g in A,

$$|\alpha(\phi)(f + ig) - \alpha(\psi)(f + ig)| = |\alpha(\overline{\phi})(f - ig) - \alpha(\overline{\psi})(f - ig)|,$$

we see that

$$\|\alpha(\phi) - \alpha(\psi)\| = \|\alpha(\overline{\phi}) - \alpha(\overline{\psi})\|.$$

Let $P_B(\alpha(\phi))$ denote the Gleason part of $\alpha(\phi)$ in $\mathrm{Car}(B)$ (see Remark 3.2.17). Then it follows that $\alpha(\psi) \in P_B(\alpha(\phi))$ if and only if $\alpha(\overline{\psi}) \in P_B(\alpha(\overline{\phi}))$. Recall that $Q_A(\phi) = \{\psi \in \mathrm{Car}(A) : \|\phi - \psi\| < 2 \text{ or } \|\overline{\phi} - \psi\| < 2\}$.

Lemma 3.3.1 For ϕ in $\mathrm{Car}(A)$,

$$\alpha(Q_A(\phi)) = P_B(\alpha(\phi)) \cup P_B(\alpha(\overline{\phi})).$$

Proof. If $\alpha(\psi) \in P_B(\alpha(\phi)) \cup P_B(\alpha(\overline{\phi}))$, then

$$\|\alpha(\phi) - \alpha(\psi)\| < 2 \qquad \text{or} \qquad \|\alpha(\overline{\phi}) - \alpha(\psi)\| < 2.$$

Hence $\|\phi - \psi\| < 2$ or $\|\overline{\phi} - \psi\| < 2$; that is, $\psi \in Q_A(\phi)$ by (ii) of Theorem 3.2.9.

Now assume that $\alpha(\psi) \notin P_B(\alpha(\phi)) \cup P_B(\alpha(\bar{\phi}))$. We shall show that $\psi \notin Q_A(\phi)$. First we assert that there is a sequence $(u_n + iv_n)$ in B such that

$$\|u_n + iv_n\| < 1,$$
$$\alpha(\phi)(u_n + iv_n) = 0 = \alpha(\phi)(u_n - iv_n) \qquad \text{for all } n,$$
$$|\alpha(\psi)(u_n + iv_n)| \longrightarrow 1 \qquad \text{as } n \longrightarrow \infty.$$

Since $\alpha(\psi) \notin P_B(\alpha(\phi))$, there is a sequence $(f_n + ig_n)$ in B such that

$$\|f_n + ig_n\| < 1,$$
$$\alpha(\phi)(f_n + ig_n) = 0 \qquad \text{for all } n,$$
$$|\alpha(\psi)(f_n + ig_n)| \longrightarrow 1 \qquad \text{as } n \longrightarrow \infty.$$

This follows from (ii) of Corollary 3.2.16. Similarly, since $\alpha(\psi) \notin P_B(\alpha(\bar{\phi}))$, there is a sequence $(h_n + ik_n)$ in B such that

$$\|h_n + ik_n\| < 1,$$
$$\alpha(\bar{\phi})(h_n + ik_n) = 0 \qquad \text{for all } n,$$
$$|\alpha(\psi)(h_n + ik_n)| \longrightarrow 1 \qquad \text{as } n \longrightarrow \infty.$$

Now let

$$u_n + iv_n = (f_n + ig_n)(h_n + ik_n).$$

Then

$$u_n - iv_n = (f_n - ig_n)(h_n - ik_n)$$

and

$$\|u_n + iv_n\| \le \|f_n + ig_n\| \, \|h_n + ik_n\| < 1.$$

Also,

$$\alpha(\phi)(u_n + iv_n) = 0$$

since

$$\alpha(\phi)(f_n + ig_n) = 0$$

and

$$\alpha(\phi)(u_n - iv_n) = 0$$

since

$$\alpha(\phi)(h_n - ik_n) = \overline{\alpha(\bar{\phi})(h_n + ik_n)} = 0.$$

Finally,

$$|\alpha(\psi)(u_n + iv_n)| = |\alpha(\psi)(f_n + ig_n)| \, |\alpha(\psi)(h_n + ik_n)| \longrightarrow 1$$

as $n \to \infty$. This proves our assertion. Now, let $s_n = \alpha(\psi)(u_n + iv_n)$. Then $|s_n| < 1$. Define

$$h_n := \frac{(u_n + iv_n - s_n)(u_n - iv_n - \bar{s}_n)}{[1 - \bar{s}_n(u_n + iv_n)][1 - s_n(u_n - iv_n)]}.$$

Then $h_n \in A$, $\|h_n\| < 1$, $\psi(h_n) = \alpha(\psi)(h_n) = 0$ for all n and

$$\phi(h_n) = \alpha(\phi)(h_n) = |s_n|^2 \longrightarrow 1 \qquad \text{as } n \longrightarrow \infty.$$

Hence $\psi \notin Q_A(\phi)$ by (iii) of Theorem 3.2.9.   □

We are now in a position to discuss the relationship between the parts of A and those of its complexification B.

Theorem 3.3.2
(a) Let $\phi, \psi \in \text{Car}(A)$. Then $\|\phi - \psi\| < 2$ if and only if $\|\alpha(\phi) - \alpha(\psi)\| < 2$. In particular, $\|\phi - \psi\| < 2$ is an equivalence relation on $\text{Car}(A)$.
(b) Let $\phi \in \text{Car}(A)$. Then

$$P_B(\alpha(\phi)) = \{\alpha(\psi) : \psi \in \text{Car}(A), \|\phi - \psi\| < 2\}.$$

Proof.
  (a) If $\|\alpha(\phi) - \alpha(\psi)\| < 2$, then

$$\|\phi - \psi\| \le \|\alpha(\phi) - \alpha(\psi)\| = 2.$$

Now let $\|\phi - \psi\| = 2$, and assume for a moment that $\|\alpha(\phi) - \alpha(\psi)\| = 2$. Then there exists a sequence $(f_n + ig_n)$ in B such that $\|f_n + ig_n\| < 1$ for all n and

$$\alpha(\phi)(f_n + ig_n) \longrightarrow 1, \qquad \alpha(\psi)(f_n + ig_n) \longrightarrow -1 \quad \text{as } n \longrightarrow \infty.$$

Since $\|\phi - \psi\| = 2$, Lemma 3.3.1 shows that

$$\alpha(\psi) \in P_B(\alpha(\phi)) \cup P_B(\alpha(\bar{\phi})).$$

But $\|\alpha(\phi) - \alpha(\psi)\| = 2$, so that

$$\alpha(\psi) \in P_B(\alpha(\bar{\phi})) \qquad \text{and} \qquad \alpha(\bar{\psi}) \in P_B(\alpha(\phi)).$$

By passing to a subsequence, we can assume by (iii) of Corollary 3.2.16 that

$$\alpha(\bar{\phi})(f_n + ig_n) \longrightarrow -1 \qquad \text{and} \qquad \alpha(\bar{\psi})(f_n + ig_n) \longrightarrow 1 \quad \text{as } n \longrightarrow \infty.$$

Thus we have

$$\phi(f_n) + i\phi(g_n) \longrightarrow 1, \qquad \psi(f_n) + i\psi(g_n) \longrightarrow -1,$$

$$\phi(f_n) - i\phi(g_n) \longrightarrow -1, \qquad \psi(f_n) - i\psi(g_n) \longrightarrow 1$$

as $n \to \infty$. Hence $\phi(g_n) \to -i$ and $\psi(g_n) \to i$ as $n \to \infty$. This contradicts $\|\phi - \psi\| < 2$, since $\|g_n\| \le \|f_n + ig_n\| < 1$ for all n. Since $\|\alpha(\phi) - \alpha(\psi)\| < 2$ is an equivalence relation on $\mathrm{Car}(B)$, we now see that $\|\phi - \psi\| < 2$ is an equivalence relation on $\mathrm{Car}(A)$.

(b) $P_B(\alpha(\phi)) = \{\alpha(\psi) : \psi \in \mathrm{Car}(A), \|\alpha(\psi) - \alpha(\phi)\| < 2\}$

$$= \{\alpha(\psi) : \psi \in \mathrm{Car}(A), \|\phi - \psi\| < 2\}, \text{ by (a).} \qquad \square$$

Corollary 3.3.3 Let ϕ in $\mathrm{Car}(A)$ be such that $\|\phi - \overline{\phi}\| = 2$. Then $\|\psi - \overline{\psi}\| = 2$ for all ψ in $Q_A(\phi)$.

Proof. Let $\psi \in Q_A(\phi)$, and assume for a moment that $\|\psi - \overline{\psi}\| < 2$. Then by (ii) of Theorem 3.2.9, $\|\phi - \psi\| < 2$ or $\|\overline{\phi} - \psi\| < 2$. If $\|\phi - \psi\| < 2$, then clearly, $\|\overline{\phi} - \overline{\psi}\| < 2$. This together with $\|\psi - \overline{\psi}\| < 2$ implies that $\|\phi - \overline{\phi}\| < 2$ by the transitivity guaranteed in Theorem 3.3.2(a). Hence $\|\phi - \psi\|$ cannot be less than 2. A similar argument shows that $\|\overline{\phi} - \psi\|$ cannot be less than 2. Thus $\|\psi - \overline{\psi}\| = 2$. $\qquad \square$

Corollary 3.3.4 Let $\phi \in \mathrm{Car}(A)$. Then
(a) $\|\phi - \overline{\phi}\| < 2$ if and only if

$$Q_A(\phi) = \{\psi \in \mathrm{Car}(A) : \|\phi - \psi\| < 2\}.$$

In this case, $P_B(\alpha(\phi)) = \alpha(Q_A(\phi)) = P_B(\alpha(\overline{\phi}))$.
(b) $\|\phi - \overline{\phi}\| = 2$ if and only if $Q_A(\phi)$ is the disjoint union of

$$\{\psi \in \mathrm{Car}(A) : \|\phi - \psi\| < 2\} \text{ and } \{\psi \in \mathrm{Car}(A) : \|\overline{\phi} - \psi\| < 2\}.$$

In this case, $\alpha(Q_A(\phi))$ is the disjoint union of $P_B(\alpha(\phi))$ and $P_B(\alpha(\overline{\phi}))$.

Proof.
(a) Let $\|\phi - \overline{\phi}\| = 2$ and $\psi \in Q_A(\phi)$. By (ii) of Theorem 3.2.9, $\|\phi - \psi\| < 2$ or $\|\overline{\phi} - \psi\| < 2$. If $\|\overline{\phi} - \psi\| < 2$, then by the transitivity guaranteed in Theorem 3.3.2(a), $\|\phi - \overline{\phi}\| < 2$ implies that $\|\phi - \psi\| = 2$. Hence

$$Q_A(\phi) = \{\psi \in \mathrm{Car}(A) : \|\phi - \psi\| < 2\}.$$

The converse is obvious since $\overline{\phi} \in Q_A(\phi)$ always.
(b) This follows by (ii) of Theorem 3.2.9, Corollary 3.3.3, and Theorem 3.3.2(a). $\qquad \square$

We have computed the Gleason parts of some real function algebras in Example 3.2.11 and the parts of their complexifications in Remark 3.2.17. In the case of the real disk algebra A, $e_0 = \overline{e}_0$. Hence

$$Q_A(e_0) = \{\psi \in \text{Car}(A) : \|\psi - e_0\| < 2\}$$

by Corollary 3.3.4. We have shown that this can be identified with the open unit disk.

The results above can be used to compute the parts of a real function algebra A if those of its complexification B are known, or to compute the parts of B if the parts of A are known. We illustrate this procedure by considering an example.

Definition 3.3.5 Let B be a complex commutative Banach algebra with unit and $\phi \in \text{Car}(B)$. Then a linear functional D on B is called a *point derivation on B at ϕ* if for all f, g in B,

$$D(fg) = D(f)\phi(g) + \phi(f)D(g).$$

(This resembles Leibnitz's rule.)

The zero functional is obviously a continuous point derivation for every ϕ. As a nontrivial example, we may consider the disk algebra B and the linear functional D defined by $D(f) = f'(0)$ for f in B. Then D is a continuous point derivation on B at e_0. For other examples and properties of point derivations, the interested reader may refer to Browder (1969, pp. 63–79).

Example 3.3.6 Let C be a complex function algebra with the maximal ideal space Z. By identifying Z with $\text{Car}(C)$ and the functions in C with their Gelfand transforms, we can regard C as a function algebra on Z (see Remark 1.2.8 and Corollary 1.2.12). Let $Z_q := \{z_1, \ldots, z_q\}$ be a specified finite subset of q points in Z, and let D_k be a continuous point derivation of C at z_k for each $k = 1, \ldots, q$. Let $A_q := \{f \in C : f(z_k) \text{ and } D_k(f) \text{ are real for } 1 \le k \le q\}$. Then A_q is a real uniform algebra. Let Y be its maximal ideal space. We now study the relationship between the Gleason parts of C and those of A_q, by first establishing a relationship between Z and Y.

Lemma 3.3.7
(i) Given a, b in Z with $a \ne b$, there exists $f_{a,b}$ in C such that $f_{a,b}(a) = 1$, $f_{a,b}(b) = 0$, and for all point derivations D_a at a and D_b at b, $D_a(f_{a,b}) = 0 = D_b(f_{a,b})$.
(ii) There exist $f_1^*, \ldots, f_q^*$ in C such that $f_m^*(z_k) = \delta_{m,k}$ and $D_k(f_m^*) = 0$ for $1 \le k, m \le q$. (Here $\delta_{m,k}$ is the usual Kronecker symbol.) In particular $f_k^* \in A_q$ for $k = 1, \ldots, q$.
(iii) For all a, b in Z with $a \ne b$, there is f in A_q such that $f(a) = 1$ and $f(b) = 0$. In particular, A_q separates the points of Z.

(iv) The restriction map $cx^*: Z \to Y$ defined by $cx^*(z) := z \cap A_q$ is a bijection.

Proof.

(i) Since C separates the points of Z, there exists h in C such that $h(a) = 1$ and $h(b) = 0$. Then $f_{a,b} := 2h^2 - h^4$ is a required function.

(ii) If $q = 1$, let $f_1^* = 1$. Now let $q \geq 2$. By (i), let for a fixed m, with $1 \leq m \leq q$,

$$f_m^* := \prod_{\substack{j=1 \\ j \neq m}}^{q} f_{z_m, z_j}.$$

It is easy to verify that for $1 \leq k \leq q\, f_m^*(z_k) = \delta_{m,k}$ and

$$D_k(f_m^*) = \sum_{\substack{j=1 \\ j \neq m}}^{q} D_k(f_{z_m, z_j}) \left(\prod_{\substack{i=1 \\ i \neq m, j}}^{q} f_{z_m, z_i} \right)(z_k) = 0.$$

(iii) If both $a, b \in Z_q$, say $a = z_m$ and $b = z_k$, let $f := f_m^*$. If $a \in Z_q$ and $b \notin Z_q$, then $a = z_m$ for some m with $1 \leq m \leq q$. Let

$$f := f_{a,b} f_m^*.$$

Clearly, $f(a) = 1$, $f(b) = 0$, and

$$D_k(f) = f_{a,b}(z_k) D_k(f_m^*) + D_k(f_{a,b}) f_m^*(z_k) = 0,$$

since $D_k(f_m^*) = 0$ by (ii) and $D_k(f_{a,b}) = 0$ by (i), if $m = k$ and $f_m^*(z_k) = 0$ by (ii) if $m \neq k$. If both a and b are not in Z_q, then let

$$h := f_{a,b} \prod_{j=1}^{q} f_{a, z_j}.$$

Since $h(a) = 1$, $h(b) = h(z_j) = 0$ for $j = 1, \ldots, q$, $f := h^2$ is a required function. This proves (iii).

(iv) Let $z \in Z$. First we prove that $cx^*(z)$ is a maximal ideal in A_q. Let $\phi \in \text{Car}(C)$ be such that $\ker \phi = z$. Let ψ be the restriction of ϕ to A_q. Clearly, $\psi \in \text{Car}(A_q)$ and $cx^*(z) = \ker(\psi)$. Hence $cx^*(z) \in Y$. Now if $a, b \in Z$ and $a \neq b$, then by (iii) above, we can find f in A_q such that $f(a) = 1$ and $f(b) = 0$. This f belongs to $cx^*(b)$, but not to $cx^*(a)$. Thus cx^* is $1:1$. To show that cx^* is onto, consider a maximal ideal M of A_q and let $I := \{fg : f \in C, g \in M\}$. Clearly, I is an ideal in C. We claim that I is proper. If not, we can find f in C and g in M such that $fg = 1$. Hence for $k = 1, \ldots, q$, $f(z_k) g(z_k) = 1$. This implies that $f(z_k)$ is real, because $g(z_k)$ is real as g belongs to A_q. Further, $0 = D_k(1) = D_k(f) g(z_k) + D_k(g) f(z_k)$. This shows that $D_k(f)$ is real as $g(z_k)$, $D_k(g)$, and $f(z_k)$ are real and $f(z_k)$ and $g(z_k)$ are nonzero. Hence $f \in A_q$. (Essentially, we have proved that if a function in A_q is invertible in C, its inverse must belong to A_q.) But then $1 = fg \in M$, a contradiction. This proves our claim. Now I is contained in a maximal ideal N of C. Then $M \subset cx^*(N)$, and hence $M = cx^*(N)$ by the maximality of M. Thus cx^* is onto. $\qquad\square$

In view of (iv) of Lemma 3.3.7, we can and shall identify Y with Z.

Theorem 3.3.8 For every z in Z, $P_{A_q}(z) = P_C(z)$.

Proof. If $z' \in P_C(z)$, then

$$\sup\{|f(z)| : f \in A_q, \quad \|f\| < 1, \quad f(z') = 0\}$$
$$\leq \sup\{|f(z)| : f \in C, \quad \|f\| < 1, \quad f(z') = 0\} < 1.$$

Hence $z' \in P_{A_q}(z)$ by (i) of Corollary 3.2.14. Thus

$$P_C(z) \subset P_{A_q}(z).$$

Now, consider $z' \notin P_C(z)$. Then by renaming the z_j's if necessary, we can assume that the first p points belong to $P_C(z)$, while $z_{p+1}, \ldots, z_q$ do not, where $0 \leq p \leq q$. Then there exist sequences $(f_{n,k})$, $p + 1 \leq k \leq q$, and a sequence (f'_n) in C such that $\|f_{n,k}\|, \|f'_n\| < 1$, $f_{n,k}(z_k) = f'_n(z') = 0$ for all n and $k = p + 1, \ldots, q$, and

$$|f_{n,k}(z)| \to 1, \quad p + 1 \leq k \leq q, \quad |f'_n(z)| \to 1 \quad \text{as } n \to \infty.$$

Let $f_n = f^2_{n,p+1} \cdots f^2_{n,q} f'_n$. Then $\|f_n\| < 1$, $f_n(z_k) = f_n(z') = 0$ for all n and $k = p + 1, \ldots, q$ and $|f_n(z)| \to 1$ as $n \to \infty$. Also,

$$D_k(f^2_n) = 2f_n(z_k)D_k(f_n) = 0$$

for all n and $k = p + 1, \ldots, q$.

Case (i): $p = 0$. Then $f^2_n \in A_q$, $\|f^2_n\| < 1$, $f^2_n(z') = 0$ for all n, and $|f^2_n(z)| \to 1$ as $n \to \infty$. Hence $z' \notin P_{A_q}(z)$.

Case (ii): $1 \leq p \leq q$. We construct functions $h_n \in A_q$ such that $\|h_n\| < 1$, $h_n(z') = 0$ for all n and $|h_n(z)| \to 1$ as $n \to \infty$. Let for $j = 1, \ldots, p$,

$$f^2_n(z_j) = \alpha_{n,j}, \quad g_{n,j} = \bar{\alpha}_{n,j} \frac{f^2_n - \alpha_{n,j}}{1 - \bar{\alpha}_{n,j} f^2_n}.$$

Let $g_n = g_{n,1} \cdots g_{n,p}$. It can be proved that $\alpha_m = g_n(z') = g_n(z_k)$, $k = p + 1, \ldots, q$, is real and

$$h_n := \frac{g^2_n - \alpha^2_n}{1 - \alpha^2_n g^2_n}$$

has the required properties, so that $z' \notin P_{A_q}(z)$. [See Kulkarni and Limaye (1981b) for details of computations.] This proves that $P_{A_q}(z) \subset P_C(z)$. Hence $P_C(z) = P_{A_q}(z)$. □

Theorem 3.3.9 Car(A_q) is homeomorphic to two copies of Z identified at Z_q. Let $\phi \in$ Car(A_q) and ker(ϕ) = z.
(i) If $P_C(z) \cap Z_q$ is empty, then $\|\phi - \overline{\phi}\| = 2$ and $Q_{A_q}(\phi)$ is homeomorphic to the disjoint union of two copies of $P_C(z)$.
(ii) If $P_C(z) \cap Z_q$ is nonempty, then $\|\phi - \overline{\phi}\| < 2$ and $Q_{A_q}(\phi)$ is homeomorphic to two copies of $P_C(z)$ identified at $P_C(z) \cap Z_q$.

Proof. We have shown in Remark 1.2.11 that the map ker: Car(A_q) $\rightarrow M(A_q) = Z$ is continuous, open, and onto. Also, for z in Z, ker$^{-1}(z) = \{\phi, \overline{\phi}\}$ for some ϕ in Car(A_q). Further, if ψ is a nonzero homomorphism on C, then the restrictions of ψ and $\overline{\psi}$ to A_q are real-linear homomorphisms on A_q. Hence

$$\text{Car}(A_q) = \{\psi|_{A_q} : \psi \in \text{Car}(C)\} \cup \{\overline{\psi}|_{A_q} : \psi \in \text{Car}(C)\}$$

and each set on the right-hand side is homeomorphic to Z. Clearly, if ker $\psi \in Z_q$, then $\psi = \overline{\psi}$. This proves the first part.

Now let $\phi \in$ Car(A_q) and ker $\phi = z$. Then $\phi(f) = f(z)$ for all f in A_q, or $\phi(f) = \overline{f(z)}$ for all f in A_q. We assume the former without loss of generality. Then by Theorem 3.3.8,

$$P_C(z) = P_{A_q}(z) = P_{A_q}(\text{ker } \phi) = \text{ker}(Q_{A_q}(\phi)).$$

(i) Let $z_j \notin P_C(z)$ for all $j = 1,...,q$. Then there exists a sequence of functions (f_n) in C such that $\|f_n\| < 1, f_n(z_j) = 0$ for all n and $j = 1,...,q$ and $f_n(z) \rightarrow 1$ as $n \rightarrow \infty$. Then it follows that for $j = 1,...,q$ and $n = 1,2,...,$

$$if_n^2(z_j) = 0 = D_j(if_n^2).$$

Hence $if_n^2 \in A_q$ for all n. Also, $\|if_n^2\| < 1$ and $if_n^2(z) \rightarrow i$. Hence $\phi(if_n^2) \rightarrow i$ and $\overline{\phi}(if_n^2) \rightarrow -i$. This shows that $\|\phi - \overline{\phi}\| = 2$. By Corollary 3.3.4, $Q_{A_q}(\phi)$ is the disjoint union of

$$\{\psi \in \text{Car}(A_q) : \|\phi - \psi\| < 2\} \quad \text{and} \quad \{\psi \in \text{Car}(A_q) : \|\overline{\phi} - \psi\| < 2\}.$$

Each of these sets is homeomorphic to $P_C(z)$ because the restriction of the map ker to each of these sets is a bijection.

(ii) Let $z_j \in P_C(z)$ for some $j = 1,...,q$. If ψ in Car(A_q) is such that $\psi(f) = f(z_j)$ for all f in A_q, then $\psi = \overline{\psi}$, so that $0 = \|\psi - \overline{\psi}\| < 2$. Since ψ is in $Q_{A_q}(\phi)$, it follows by Corollary 3.3.3 that $\|\phi - \overline{\phi}\| < 2$. Hence by Corollary 3.3.4,

$$Q_{A_q}(\phi) = \{\psi \in \mathrm{Car}(A_q) : \|\phi - \psi\| < 2\}.$$

In this case, the map ker: $Q_{A_q}(\phi) \to P_C(z)$ is at most $2:1$, that is, $\ker(\psi) = \ker(\overline{\psi})$ for all ψ, and $\psi = \overline{\psi}$ if and only if $\ker \psi$ belongs to Z_q. This proves (ii). ☐

Corollary 3.3.10 Let $B = \{f + ig : f, g \in A_q\}$. Then $\mathrm{Car}(B)$ is homeomorphic to the space obtained from two copies of Z identified at Z_q. Let $\psi \in \mathrm{Car}(B)$ and $z = \ker(\psi_{|A_q})$.
(i) If $P_C(z) \cap Z_q$ is empty, then $P_B(\psi)$ is homeomorphic to $P_C(z)$.
(ii) If $P_C(z) \cap Z_q$ is nonempty, then $P_B(\psi)$ is homeomorphic to two copies of $P_C(z)$ identified at $P_C(z) \cap Z_q$.

Proof. The first part follows by observing that the map α is a homeomorphism from $\mathrm{Car}(A_q)$ to $\mathrm{Car}(B)$. Now let $\phi = \psi_{|A_q}$. Then by Theorem 3.3.2,

$$P_B(\psi) = P_B(\alpha(\phi))$$
$$= \{\alpha(\theta) : \theta \in \mathrm{Car}(A_q), \quad \|\phi - \theta\| < 2\}.$$

Case (i): $\|\phi - \overline{\phi}\| = 2$ by Theorem 3.3.9(i). It is also shown in this proof that the set $\{\theta \in \mathrm{Car}(A_q) : \|\theta - \phi\| < 2\}$ is homeomorphic to $P_C(z)$.

Case (ii): $\|\phi - \overline{\phi}\| < 2$ by Theorem 3.3.9(ii). Then $P_B(\psi) = P_B(\alpha(\phi)) = \alpha(Q_{A_q}(\phi))$ by Corollary 3.3.4 and $Q_{A_q}(\phi)$ is homeomorphic to two copies of $P_C(z)$ identified at $P_C(z) \cap z_q$ by Theorem 3.3.9(ii). ☐

The algebra A_q was defined in Limaye and Simha (1975). This paper also contains a proof of Lemma 3.3.7. Other results in this section were proved in Kulkarni and Limaye (1981b).

3.4 TOPOLOGICAL CHARACTERIZATION OF GLEASON PARTS

We have observed in Remark 3.2.13 that if Q is a Gleason part of a real function algebra A, then Q is completely regular and σ-compact, and the map $\phi \mapsto \overline{\phi}$ is a topological involution on Q. This raises the following natural question. Suppose that K is a completely regular σ-compact topological space with a topological involution τ on it. Then can K be viewed as a Gleason part of a real function algebra A, so that τ is the restriction to K of the natural involution $\phi \mapsto \overline{\phi}$ on $\mathrm{Car}(A)$? We answer this question in the affirmative in this section. In the case of a complex function algebra, Garnett (1967) proved that every completely regular σ-compact topological space can be looked upon as a Gleason part of some complex function algebra. We shall use many ideas of Garnett in proving our assertion. We begin with the discussion of an example.

Example 3.4.1 (Real big disk algebra) Let T^2 be the torus defined by

$$T^2 := \{s = (s_1, s_2) \in \mathbb{C}^2 : |s_1| = |s_2| = 1\}.$$

For $s = (s_1, s_2)$ in T^2, let $\tau(s) := \bar{s} := (\bar{s}_1, \bar{s}_2)$. Let z_1 and z_2 be the "coordinate functions" $z_1(s) := s_1$ and $z_2(s) := s_2$ for all s in T^2. Let β be any positive irrational number, and let A_β be the real closed subalgebra of $C(T^2)$ generated by the functions $\{z_1^n z_2^m : n, m \text{ integers}, n + m\beta \geq 0\}$. Then A_β is a real function algebra on (T^2, τ). A_β will be called the *real big disk algebra*.

The complex closed subalgebra B_β of $C(T^2)$ generated by the same set of functions is called the *big disk algebra*. Obviously,

$$B_\beta = \{f + ig : f, g \in A_\beta\}.$$

For a detailed treatment of and further references about the big disk algebra, we refer the reader to Browder (1969, p. 44). We now examine the carrier space of the real big disk algebra A_β. In particular, we show that the carrier space can be identified with the set

$$L := \{(s_1, s_2) \in \mathbb{C}^2 : |s_1| \leq 1, \quad |s_2| = |s_1|^\beta\}.$$

Let $f \in \text{Car}(A_\beta)$ and $\phi(z_1) = s_1$, $\phi(z_2) = s_2$. Then $|s_1| \leq 1$ and $|s_2| \leq 1$. Also, for all n and m with $n + m\beta \geq 0$, we have

$$|\phi(z_1^n z_2^m)| = |s_1^n s_2^m| \leq 1,$$

that is

$$n \log|s_1| + m \log|s_2| \leq 0.$$

Since this inequality is satisfied for all n and m with $n + m\beta \geq 0$, and β is irrational, we have $\log |s_2| = \beta \log|s_1|$, so that $|s_2| = |s_1|^\beta$. Since ϕ is clearly determined by (s_1, s_2), we get a $1:1$ continuous map of $\text{Car}(A_\beta)$ into L. To show that this map is onto, consider (s_1, s_2) in L such that $(s_1, s_2) \neq (0,0)$. Let p be a finite sum of the form $\Sigma c_{nm} z_1^n z_2^m$, $n + m\beta \geq 0$. We define $\phi(p) := \Sigma c_{nm} s_1^n s_2^m$. Since $|s_2| = |s_1|^\beta$, there exists a real number t such that $s_2 = \exp(it)s_1^\beta$. Further, since $s_1 \neq 0$, we can find w such that $\text{Re } w \leq 0$ and $s_1 = \exp(w)$. Then $s_2 = \exp(it + bw)$ and $\phi(p) = \Sigma c_{nm} \exp(imt) \exp((n + m\beta)w) =: g(w)$, say. Then $g(w)$ is analytic on $\{w \in \mathbb{C} : \text{Re } w < 0\}$ and continuous and bounded on $\{w \in \mathbb{C} : \text{Re } w \leq 0\}$. Hence by the Phragmen–Lindeloff principle,

$$|g(w)| \leq \sup\{|g(ix)|:x \text{ real}\}.$$

But for real x,

$$g(ix) = \sum c_{nm} \exp(inx) \exp(im(\beta x + t))$$

is the value of p at the point $(\exp(ix), \exp(i(\beta x + t)))$. Thus $|\phi(p)| \leq \|p\|$. Hence ϕ extends uniquely to a nonzero continuous linear map from A_β to $\mathbb{C}$. Clearly, ϕ and hence its extension are multiplicative. Thus the image of $\text{Car}(A_\beta)$ under the continuous map $\phi \mapsto (\phi(z_1), \phi(z_2))$ contains $L \backslash \{(0,0)\}$ and hence L. [The same proof also shows that $\text{Car}(B_\beta)$ can also be identified with L.]

Also, if the image of ϕ under the map above is (s_1, s_2), then that of $\bar\phi$ is $(\bar{s}_1, \bar{s}_2)$. Thus the map τ defined on T^2 has a natural extension to L. We shall denote this extension also by τ. A_β has a one-point part $\{(0,0)\}$, which is called the *origin of the big disk*. This can be seen as follows. Let ψ_0 denote the "evaluation" at $(0,0)$, and ϕ the evaluation at $s := (s_1, s_2)$ in $L \backslash \{(0,0)\}$. Since β is irrational, we can find integers n_k and m_k such that $n_k + m_k\beta \geq 0$ for all k and $n_k + m_k\beta \to 0$ as $k \to \infty$. Let $f_k := z_1^{n_k} z_2^{m_k}$. Then $f_k \in A_\beta$, $\|f_k\| \leq 1$, $\psi_0(f_k) = 1$ for all k, and

$$\begin{aligned}|\phi(f_k)| &= |s_1|^{n_k} |s_2|^{m_k} \\ &= |s_1|^{(n_k + m_k\beta)} \to 1 \qquad \text{as } k \to \infty.\end{aligned}$$

Thus $\{\psi_0\}$ is a one-point part. Note that $\psi_0 = \bar\psi_0$. In the sequel we shall also use the fact that the normalized Haar measure on T^2 is a RPR measure for $\text{Re}(\psi_0)$. Note that since $\psi_0 = \bar\psi_0$, we have $\text{Re } \psi_0 = \psi_0$.

Lemma 3.4.2 Let A be a real commutative Banach algebra with unit, Y a nonempty closed subset of $\text{Car}(A)$, and B the uniform closure of $\{\hat{f}_{|Y} : f \in A\}$. Then B is a real commutative Banach algebra and $\text{Car}(B)$ can be identified with the following subset of $\text{Car}(A)$:

$$Z := \{\phi \in \text{Car}(A) : |\phi(f)| \leq \|\hat{f}\|_Y \text{ for all } f \text{ in } A\}.$$

Proof. B is clearly a real commutative Banach algebra. [Recall that $\|\hat{f}\|_Y = \sup\{|\phi(f)| : \phi \in Y\}$.] For ϕ in Z and f in A, we define ϕ' by $\phi'(\hat{f}_{|Y}) = \phi(f)$. Then ϕ' is well-defined: If $\hat{f}_{|Y} = \hat{g}_{|Y}$ for f, g in A, then $|\phi(f) - \phi(g)| \leq \|\hat{f} - \hat{g}\|_Y = 0$. Also, $|\phi'(\hat{f}_{|Y})| = |\phi(f)| \leq \|f\|$. Hence ϕ' can be uniquely extended as a continuous linear map to B. We denote this extension also by ϕ'. Clearly, ϕ' is nonzero and multiplicative.

Now let $\psi \in \text{Car}(B)$. We shall show that $\psi = \phi'$ for some ϕ in Z. Let

$$I := \{f \in A : \psi(\hat{f}_{|Y}) = 0\}.$$

Clearly, I is a proper ideal of A and is hence contained in a maximal ideal M of A. By Theorem 1.2.6, $M = \ker \phi$ for some ϕ in $\mathrm{Car}(A)$. Now define $\phi'(\hat{f}_{|Y}) = \phi(f)$ for f in A. Then ϕ' is well-defined, because if $\hat{f}_{|Y} = \hat{g}_{|Y}$ for some f, g in A, then $f - g \in I \subset M$. Hence $\phi(f) = \phi(g)$. Thus, as above, $\phi' \in \mathrm{Car}(B)$. But $\ker \psi \subset \ker \phi'$; hence $\ker \psi = \ker \phi'$ by the maximality of $\ker \psi$. Again by Theorem 1.2.6, $\psi = \phi'$ or $\psi = \overline{\phi'} = \overline{\phi}'$. Further, for f in A, $|\phi(f)| = |\phi'(\hat{f}_{|Y})| \leq \|\hat{f}\|_Y$. Thus ϕ as well as $\overline{\phi}$ belong to Z. $\square$

Definition 3.4.3 Let A be a real or complex commutative Banach algebra with unit. For a subset E of $\mathrm{Car}(A)$, the *kernel of E*, is a subset $\ker(E)$ of A, defined by

$$\ker(E) := \cap \{\ker \phi : \phi \in E\}.$$

For a subset J of A, the *hull of J* is a subset $\mathrm{hull}(J)$ of $\mathrm{Car}(A)$ defined by

$$\mathrm{hull}(J) := \{\phi \text{ in } \mathrm{Car}(A) : J \subset \ker \phi\}.$$

A subset E of $\mathrm{Car}(A)$ is called *hull-kernel closed* if $\mathrm{hull}(\ker(E)) = E$.

It follows from the definition above that for a subset E of $\mathrm{Car}(A)$, $\phi_0 \in \mathrm{Car}(A) \backslash \mathrm{hull}(\ker(E))$ if and only if there exists a in A such that $\phi(a) = 0$ for all ϕ in E and $\phi_0(a) \neq 0$. The mapping $E \mapsto \mathrm{hull}(\ker(E))$ is a Kuratowski closure operation on the subsets of $\mathrm{Car}(A)$, leading to a topology on $\mathrm{Car}(A)$ called the *hull-kernel topology*. For further information about this topology, we refer the reader to Bonsall and Duncan (1973, p. 115).

Theorem 3.4.4 Let A be a real function algebra on $(\mathrm{Car}(A), \tau)$, where $\tau(\phi) = \overline{\phi}$ for ϕ in $\mathrm{Car}(A)$. Let Q be a Gleason part of A and S a hull-kernel closed subset of $\mathrm{Car}(A)$ such that $\tau(S) = S$. Then there exists a real function algebra A', a Gleason part Q' of A', and a homeomorphism F of $Q \cap S$ onto Q' such that $F(\overline{\phi}) = \overline{F(\phi)}$ for all ϕ in $Q \cap S$, and $\|F(\phi_1) - F(\phi_2)\| \leq \|\phi_1 - \phi_2\|$ for all ϕ_1 and ϕ_2 in $Q \cap S$.

Proof. All the function algebras in this proof are taken to be on their carrier spaces with natural involutions. Hence we shall identify functions with their Gelfand transforms.

Let A_β be the real big disk algebra of Example 3.4.1. Let $A_\beta \otimes A$ be the real function algebra on $\mathrm{Car}(A_\beta) \times \mathrm{Car}(A)$ generated by the functions of the form $(f \otimes g)(x,y) = f(x)g(y)$ for f in A_β, g in A, x in $\mathrm{Car}(A_\beta)$, and y in $\mathrm{Car}(A)$. Then $\mathrm{Car}(A_\beta \otimes A)$ is homeomorphic to $\mathrm{Car}(A_\beta) \times \mathrm{Car}(A)$ in a natural manner. Let

$$J := \{(s_1, s_2) \in T^2 : \mathrm{Re}\, s_j \leq 0, \quad j = 1,2\}$$
$$X := (J \times \mathrm{Car}(A)) \cup (\mathrm{Car}(A_\beta) \times S).$$

Let A' be the uniform closure of $\{h_{|X} : h \in A_\beta \otimes A\}$. Then by Lemma 3.4.2, A' can be regarded as a real function algebra on $\mathrm{Car}(A')$, which can be identified with the following set:

$$Z := \{q \in \mathrm{Car}(A_\beta \otimes A) : |g(q)| \le \|g\|_X \text{ for all } g \text{ in } A_\beta \otimes A\}.$$

Claim 1. $Z = X$. Clearly, $X \subset Z$. Now suppose that $(x^0, y^0) \in \mathrm{Car}(A_\beta \otimes A) \setminus X$. Then $x^0 \notin J$ and $y^0 \notin S$. Since S is hull-kernel closed, there exists g in A with $g(y^0) = 1$ and $g(y) = 0$ for all y in S. Let $x^0 = (t_1, t_2) \notin J$. Then $\mathrm{Re}\, t_1 > 0$ or $\mathrm{Re}\, t_2 > 0$. Without loss of generality we may assume that $\mathrm{Re}\, t_1 > 0$. If z_1 denotes the first coordinate function (as in Example 3.4.1), let $f = \exp(-\mathrm{Re}\, t_1) \exp(z_1)$. For $s = (s_1, s_2)$ in J, $|\exp(s_1)| = \exp(\mathrm{Re}\, s_1) \le 1$. Hence $\|f\|_J < \exp(-\mathrm{Re}\, t_1) < 1$, whereas $|f(x^0)| = 1$. We can choose n such that $\|f^n\|_J < 1/\|g\|$. Then $h := f^n \otimes g \in A_\beta \otimes A$, $|h(x^0, y^0)| = 1$ and $\|h\|_X < 1$. Thus $(x^0, y^0) \notin Z$. This proves the claim.

Now let ψ_0 be as in Example 3.4.1. We have seen there that $\{\psi_0\}$ is a one-point part of A_β, $\psi_0 = \overline{\psi}_0$, and the normalized Haar measure on T^2 is a RPR measure for $\mathrm{Re}\, \psi_0 = \psi_0$. Let $Q' = \{\psi_0\} \times (Q \cap S)$. Then $Q' \subset X$.

Claim 2. Q' is a Gleason part of A' in $\mathrm{Car}(A')$. We define $F: Q \cap S \to Q'$ by $F(\phi) = (\psi_0, \phi)$ for ϕ in $Q \cap S$. Note that $\overline{F(\phi)} = (\overline{\psi}_0, \overline{\phi}) = (\psi_0, \overline{\phi}) = F(\overline{\phi})$ for all ϕ in $Q \cap S$.

Now let $\phi_1, \phi_2 \in Q \cap S$. Let $h \in A'$ with $\|h\| < 1$. For ϕ in $\mathrm{Car}(A)$, define $g(\phi) := h(\psi_0, \phi)$. Then $g \in A$ and $\|g\| \le 1$. (Recall that we identify functions with their Gelfand transforms.) Now

$$|h(F(\phi_1) - h(F(\phi_2))| = |h(\psi_0, \phi_1) - h(\psi_0, \phi_2)|$$
$$= |g(\phi_1) - g(\phi_2)|.$$

Thus $\|F(\phi_1) - F(\phi_2)\| \le \|\phi_1 - \phi_2\|$. Similarly, $\|F(\phi_1) - F(\overline{\phi}_2)\| \le \|\phi_1 - \overline{\phi}_2\|$. [Note that $\overline{\phi}_2 \in Q \cap S$ as $\tau(S) = S$ and $\tau(Q) = Q$.] Now by Theorem 3.2.9(ii), $\|\phi_1 - \phi_2\| < 2$ or $\|\phi_1 - \overline{\phi}_2\| < 2$. Hence $F(\phi_1)$ and $F(\phi_2)$ belong to the same Gleason part of A'.

Let $(\psi_0, \phi_1) \in Q'$ and $(\psi, \phi_2) \in X \setminus Q'$. If $\psi \ne \psi_0$, then since $\{\psi_0\}$ is a Gleason part of A_β, there exists a sequence f_n in A_β such that $\|f_n\| < 1$, $f_n(\psi) = 0$ for all n and $|f_n(\psi_0)| \to 1$ as $n \to \infty$ by Theorem 3.2.9(iii). But then $h_n = (f_n \otimes 1)_{|X} \in A'$, $\|h_n\| < 1$, $h_n(\psi, \phi_2) = f_n(\psi) = 0$ for all n and $|h_n(\psi_0, \phi_1)| = |f_n(\psi_0)| \to 1$ as $n \to \infty$. Hence by Theorem 3.2.9(iii), (ψ_0, ϕ_1) and (ψ, ϕ_2) belong to different parts of A'. If $\psi = \psi_0$, then $\phi_2 \notin Q \cap S$. If $\phi_2 \notin S$, then by the definition of X, $\psi \ne \psi_0$. Hence we must have $\phi_2 \notin Q$. Then, as above, we can choose g_n in A, with $\|g_n\| < 1$, $g_n(\phi_2) = 0$ for all n and $|g_n(\phi_1)| \to 1$ as $n \to \infty$. By considering $h_n = (1 \otimes g_n)_{|X}$ in A', we conclude that (ψ_0, ϕ_1) and (ψ, ϕ_2) belong to different parts of A'. This proves the claim.

Clearly, the map F is a homeomorphism of $Q \cap S$ onto Q'. Other properties of F are already proved in the course of this proof.   □

Remark 3.4.5 For a real function algebra A and ϕ in Car(A), we have already seen that $\overline{\phi}$ always belongs to $Q_A(\phi)$. It is possible that $Q_A(\phi)$ may contain only ϕ and $\overline{\phi}$. For example, if A is the real disk algebra, we have shown in Example 3.2.11 that if ϕ is the evaluation at $\exp(i\theta)$, $0 < \theta < \pi$, then $Q_A(\phi) = \{\phi, \overline{\phi}\}$ and $\|\phi - \overline{\phi}\| = 2$. This raises the following question: Does there exist a Gleason part Q of a real function algebra A such that $Q = \{\phi, \overline{\phi}\}$ and $\|\phi - \overline{\phi}\| < 2$? Theorem 3.4.4 can be used to answer this question affirmatively. Let A be the real disk algebra, ϕ be the evaluation at $i/2$, and $S := \{\phi, \overline{\phi}\}$. The function $1 + 4z^2$ can be used to show that S is hull-kernel closed. Let $Q := \{e_s : |s| < 1\}$. Then Q is a Gleason part of A. Now since $\tau(S) = S$, by Theorem 3.4.4, there exists a real function algebra A', a part Q' of A', and a homeomorphism F of $Q \cap S = S$ onto Q'. Further,

$$\|F(\phi) - F(\overline{\phi})\| \leq \|\phi - \overline{\phi}\| < 2.$$

(That $\|\phi - \overline{\phi}\| < 2$ was proved in Example 3.2.11.)

Example 3.4.6 Let I be an index set, Y be the closed unit disk in the complex plane, $Y_i = Y$ for i in I, and Y_I be the product $\Pi_{i \in I} Y_i$. For $y = \{y(i), i \in I\}$, we define $\tau(y) := \overline{y} := \{\overline{y(i)}, i \in I\}$. Y_I is a compact Hausdorff space and τ is a topological involution on Y_I. Let A_I be the *real* closed subalgebra of $C(Y_I)$ generated by the coordinate function z_i, i in I, where $z_i(p) = p(i)$ for p in Y_I. Then A_I is a real function algebra on (Y_I, τ). For ϕ in Car(A_I), we have $|\phi(z_i)| \leq 1$, so that ϕ is the evaluation map at λ in Y_I, where $\lambda(i) = \phi(z_i)$ for i in I. Thus we can identify Car(A_I) with Y_I. Let θ, with $\theta(i) = 0$ for all $i \in I$, be the "origin" in Y_I and Q_0 be the Gleason part in Car(A_I)($= Y_I$) containing θ. Let $p \in$ Car(A_I). Then using (iii) and (v) of Theorem 3.2.9, it follows that $p \in Q_0$ if and only if there exists β, $0 < \beta < 1$, such that $|z_i(p)| < \beta$ for all i in I.

Theorem 3.4.7 Let K be a completely regular σ-compact topological space with a topological involution τ on it. Then there is a real function algebra A, a Gleason part Q of Car(A), and a homeomorphism F of K onto Q such that $F(\tau(x)) = \overline{F(x)}$ for every x in K.

Proof. Let

$$K' = \begin{cases} K \cup \{\infty\}, & \text{if } K \text{ is compact} \\ \beta(K), \text{ (Stone Čech compactification of } K) & \text{if } K \text{ is noncompact.} \end{cases}$$

Then K' is a compact Hausdorff space and τ can be extended uniquely to an involution on K'. Let

$$I = K' \backslash K \qquad (I \text{ is a singleton if } K \text{ is compact})$$

and

$$V := \{f \in C(K' \times Y_I) : f_{|\{x\} \times Y_I} \in A_I$$

$$\text{for all } x \text{ in } K' \text{ and } f|_{K' \times \{\theta\}} \text{ is constant.}\}$$

Then $\mathrm{Car}(V)$ can be identified with $K' \times Y_I / \approx$, where $\approx$ identifies $K' \times \{\theta\}$ to a point. Let τ' be the natural involution $\phi \to \bar{\phi}$ on $\mathrm{Car}(V)$. If ϕ is (the evaluation map at) the point (x,y), x in K', y in Y_I, then $\tau'(\phi) = \bar{\phi}$ is (the evaluation map at) $(\tau(x),\bar{y})$, where $\bar{y}$ is as defined in Example 3.4.6. V is a real function algebra on $(\mathrm{Car}(V),\tau')$. Also,

$$Q := \{(x,y) \in \mathrm{Car}(V) : y \in Q_0\},$$

where Q_0, as defined in Example 3.4.6, is a Gleason part of A_I in $\mathrm{Car}(A_I)$. Let (x_1,y_1), (x_2,y_2) be in Q. Then since $f_{|\{x\} \times Y_I}$ is a function in A_I for f in V, $(x,y) \sim (x,\theta)$ whenever y is in Q_0. Thus $(x_1,y_1) \sim (z_1,\theta)$, $(x_2,\theta) \sim (x_2,y_2)$, and (z_1,θ) is identified with (x_2,θ). Hence $(x_1,y_1) \sim (x_2,y_2)$ and Q is a Gleason part of V.

Now we show that there exists a hull-kernel closed set S in $\mathrm{Car}(V)$ such that $\tau'(S) = S$ and a homeomorphism G of K onto $Q \cap S$ such that $G(\tau(x)) = \tau'(G(x))$ for all x in K. Then an obvious application of Theorem 3.4.4 would complete our proof.

Case (i): Let K be compact, so that $K' = K \cup \{\infty\}$. Since $\{\infty\}$ is isolated in K', there is a continuous function $h: K' \to [\frac{1}{2},1]$ such that $h^{-1}(1) = \{\infty\}$ and $h(\tau(x)) = h(x)$ for all x in K. Let

$$S := \{(x,h(x)) : x \in K'\} \subset \mathrm{Car}(V).$$

Since I is a singleton, there is only one coordinate function z_i in A_I, and we denote this by z. Then the function $g(x,z) := (h(x) - z)/(3h(x) - z)$ vanishes exactly on S and belongs to V. Hence S is hull-kernel closed. Now for $(x,h(x))$ in S,

$$\tau'(x,h(x)) = (\tau(x),\bar{h}(x)) = (\tau(x),h(x))$$
$$= (\tau(x),h(\tau(x)) \in S.$$

Thus $\tau'(S) = S$.

Let $G: K \to Q \cap S$ be defined by $G(x) := (x,h(x))$. Then G is a homeomorphism of K onto $Q \cap S$ and for x in K,

$$G(\tau(x)) = (\tau(x),h(\tau(x)))$$
$$= \tau'(x,h(x))$$
$$= \tau'(G(x)).$$

Case (ii): Let K be noncompact, so that $K' = \beta(K)$. We can write $K = \bigcup_{n=1}^{\infty} K_n$, with $K_n \subset K_{n+1}$ for all n and each K_n compact. Then for each t in $K' \setminus K$, there exists a continuous function $h_t: K' \to [\frac{1}{2}, 1]$ with $h_t(t) = 1$, $h_t(x) = 1 - 2^{-n}$ for all x in K_n, and $h_t(\tau(x)) = h_t(x)$ for all x in K. Let G: $\beta(K) \to \mathrm{Car}(V)$ be defined by $G(x) = (x, H(x))$, where $H(x)(t) = h_t(x)$ for each t in I. Then G is a homeomorphism of $\beta(K)$ onto $G(\beta(K)) := S$ and $G(K) = Q \cap S$. Also, as in case (i), it follows that $\tau'(S) = S$, and $G(\tau(x)) = \tau'(G(x))$ for all x in K. Finally, S is hull-kernel closed in $\mathrm{Car}(V)$ because

$$S = \cap \{ g_t^{-1}(0) : t \in I \},$$

where

$$g_t(x, z) := \frac{h_t(x) - z_t}{3h_t(x) - z_t}, \qquad t \in I. \qquad \square$$

Theorem 3.4.8 Let K be a σ-compact completely regular topological space. Then there exists a complex function algebra B and a Gleason part P in $\mathrm{Car}(B)$ [$= M(B)$] such that K is homeomorphic to P.

Proof. Let τ be the identity involution on K. By Theorem 3.4.7, there exists a real function algebra A, a part Q in $\mathrm{Car}(A)$ and a homeomorphism F of K onto Q such that $F(x) = \overline{F(x)}$, for all x in K. Let $B := \{f + ig : f, g \in A\}$. Now by Corollary 3.3.4(b), for x in K, $\alpha(Q) = \alpha(Q_A(F(x))) = P_B(\alpha(F(x)))$ as $F(x) = \overline{F(x)}$. Thus $\alpha \circ F$ is a homeomorphism of K onto $P_B(\alpha(F(x)))$. $\qquad \square$

Theorem 3.4.8 was proved in Garnett (1967). Kulkarni and Limaye (1983) used Garnett's result and the complexification technique to prove Theorem 3.4.7. On the other hand, the proof of Theorem 3.4.7 given above is intrinsic in nature and contains Garnett's results as a special case.

3.5 PARTS, ANALYTICITY, AND HARMONICITY

Let X be a compact Hausdorff space, τ be a topological involution on X, and A be a real function algebra on (X, τ). Let $B := \{f + ig : f, g \in A\}$ be a complexification of A. In this section we apply a well-known result about complex function algebras due to Wermer, Hoffman, Lumer, and Gamelin [see, e.g., p. 161 of Gamelin (1969)] to the algebra B and use it to deduce the existence of an analytic structure in $\mathrm{Car}(A)$. This, in turn, implies the presence of harmonic structure in $M(A)$ in the form of a connected finite Klein surface. This section heavily uses concepts appearing in the monograph by Alling and Greenleaf (1971). The section is at a slightly more advanced level than the rest of the chapter. The material appearing in this section is not used subsequently.

Theorem 3.5.1 Let A be a real function algebra on (X, τ) and $\phi \in$ Car(A). Suppose that there is $\psi \neq \phi$ in Car(A) such that $\|\phi - \psi\| < 2$. Also, assume that the linear span of the set of all regular Borel probability measures μ on X satisfying $\int_X f\, d\mu = \phi(f)$ for all f in A is finite dimensional and that there exists a unique regular Borel probability measure v on X satisfying

$$\log |\phi(f) + i\phi(g)| = \int_X \log|f + ig|\, dv$$

for all pairs of functions f, g in A with $f^2 + g^2$ invertible in A. Let

$$W := \{\theta \in \text{Car}(A) : \|\theta - \phi\| < 2\}$$

and

$$\overline{W} := \{\theta \in \text{Car}(A) : \|\theta - \overline{\phi}\| < 2\}.$$

Then W and $\overline{W}$ can be given the structures of connected finite open Riemann surfaces in such a way that for every f in A, $\hat{f}$ is a bounded analytic function on W as well as on $\overline{W}$. Moreover, with respect to these structures, the map $\tau_0 : W \to \overline{W}$ given by $\tau_0(\theta) = \overline{\theta}$ is antianalytic.

In particular, if there is a unique probability measure μ on X satisfying $\int_X f\, d\mu = \phi(f)$ for all f in A, then W and $\overline{W}$ can, in fact, be given the structure of an open unit disk in $\mathbb{C}$.

Proof. Since $\|\phi - \psi\| < 2$, $P = P_B(\alpha(\phi))$ is nontrivial by Theorem 3.3.2(b). The assumed conditions imply that the set of representing measures for $\alpha(\phi)$ is finite dimensional and $\alpha(\phi)$ has a unique log-modular measure. [For definitions, see pp. 31 and 110 of Gamelin (1969).] Hence by Theorem 7.5, Chapter VI of Gamelin (1969), P can be given the structure of a connected finite open Riemann surface such that $(f + ig)^{\wedge}$ is a bounded analytic function for every $f + ig$ in B. Hence $W := \alpha^{-1}(P)$ can be given the structure of a connected finite open Riemann surface such that for every f in A, $\hat{f}$ is a bounded holomorphic function on W.

Note that for any $\alpha(\psi)$ in Car(B), v is a representing (respectively, log-modular) measure for $\alpha(\psi)$ if and only if v' is a representing (respectively, log-modular) measure for $\alpha(\overline{\psi})$, where v' is defined by

$$v'(E) := v(\{\alpha(\overline{\theta}) : \alpha(\theta) \in E\}).$$

This shows that $\alpha(\overline{\phi})$ also has a unique log-modular measure and the dimension of the set of representing measures for $\alpha(\overline{\phi})$ is the same as that for $\alpha(\phi)$; hence it is finite. Thus $\overline{W} = \{\theta \in \mathrm{Car}(A) : \|\theta - \overline{\phi}\| < 2\}$ can be given the structure of a finite open Riemann surface in exactly the same fashion as above.

To prove that the map $\tau_0: W \to \overline{W}$ is antianalytic, let $U = (U_j, \alpha_j)_{j \in J}$ and $V = (V_k, \beta_k)_{k \in K}$ be analytic atlases over W and $\overline{W}$. [For a definition, see p. 5 of Alling and Greenleaf (1971).] Let $\theta \in W$ be such that $\theta \in U_j$ and $\tau_0(\theta) \in V_k$. We can find a bounded analytic function F on $\overline{W}$ whose ramification index at $\tau_0(\theta)$ is 1. [For a definition, see p. 27 of Alling and Greenleaf (1971).] By Theorem 7.5, Chapter VI of Gamelin (1969), there exists a sequence $(f_n + ig_n)$ in B such that $(f_n + ig_n)\hat{}$ converges to F uniformly on compact subsets of $\overline{W}$. Since $\hat{f}_n$ and $\hat{g}_n$ are bounded analytic functions on $\overline{W}$, $(f_n + ig_n)\hat{} \circ \tau_0 = \hat{f}_n + i\hat{g}_n$ is antianalytic on W for each n. Hence $F \circ \tau_0$ is antianalytic.

Now, let

$$f := \beta_k \circ \tau_0 \circ \alpha_j^{-1}, \qquad g := F \circ \beta_k^{-1}, \qquad h := F \circ \tau_0 \circ \alpha_j^{-1} = g \circ f.$$

Then g is analytic and h is antianalytic. Let $w := f(z)$. We have

$$\frac{\partial h}{\partial z} = \frac{\partial g}{\partial w} \frac{\partial f}{\partial z} + \frac{\partial g}{\partial \overline{w}} \frac{\partial f}{\partial z}$$

by Lemma 1.1.2 of Alling and Greenleaf (1971). Since h is antianalytic, $\partial h/\partial z = 0$, and since g is analytic $\partial g/\partial \overline{w} = 0$. Thus

$$\frac{\partial g}{\partial w} \frac{\partial f}{\partial z} = 0.$$

But $\partial g/\partial w \neq 0$ as the ramification index of F at $\tau_0(\theta)$ is 1. Hence $\partial f/\partial z = 0$; that is, $\beta_k \circ \tau_0 \circ \alpha_j^{-1}$ is antianalytic. Hence τ_0 is antianalytic.

Finally, if $\alpha(\phi)$ has a unique representing measure on X, then $P = P_B(\alpha(\phi))$ can be given the structure of an open unit disk [see Theorem 7.2, Chapter VI of Gamelin (1969)]. Hence W and $\overline{W}$ can also be given the structure of an open unit disk.   □

Remark 3.5.2 It follows from the discussion in Example 3.2.11 that the nontrivial Gleason parts of the algebras in Examples 1.3.10, 1.3.11, and 1.3.12 have analytic structures. Limaye and Simha (1975) have given further examples of algebras that satisfy the conditions of Theorem 3.5.1.

Corollary 3.5.3 Assume that the hypotheses in Theorem 3.5.1 hold.
(i) If $\|\phi - \overline{\phi}\| = 2$, then $Q_A(\phi)$ is the disjoint union of the two connected finite open Riemann surfaces W and $\overline{W}$. Also, $V = P_A(\ker(\phi)) \subset M(A)$

can be given the structure of a connected finite open Riemann surface (without boundary) in such a way that $\operatorname{Re} \hat{f}$ is a bounded harmonic function on V for every f in A.

(ii) If $\|\phi - \overline{\phi}\| < 2$, then $V = P_A(\ker(\phi)) \subset M(A)$ can be given the structure of a connected finite Klein surface (without boundary) in such a way that for every f in A, $\operatorname{Re} \hat{f}$ is a bounded harmonic function on V. [For a definition, see p. 6 of Alling and Greenleaf (1971).] W is canonically isomorphic to the complex double V_c of V. [For a definition, see p. 40 of Alling and Greenleaf (1971).] If τ_0 has no fixed points in W (i.e., there is no $\psi \in W$ such that $\psi = \overline{\psi}$), then V is nonorientable and $W = V_c$ is also the orienting double V_0 of V.

Proof.

(i) If $\|\phi - \overline{\phi}\| = 2$, then by Corollary 3.3.4(b), $Q_A(\phi)$ is the disjoint union of the connected finite open Riemann surfaces

$$W = \{\theta \in \operatorname{Car}(A) : \|\theta - \phi\| < 2\}$$

and

$$\overline{W} = \{\theta \in \operatorname{Car}(A) : \|\theta - \overline{\phi}\| < 2\}.$$

Since $\ker_{|W}$ is $1:1$ and onto $V = P_A(\ker(\phi))$, V is also a connected finite open Riemann surface.

(ii) Let now $\|\phi - \overline{\phi}\| < 2$. Then by Corollary 3.3.4(a), $W = \overline{W}$. Now τ_0 is an antianalytic involution on W. The quotient space W/τ_0 can be identified with V via the quotient map ker. By Theorem 1.8.4 of Alling and Greenleaf (1971), V has a unique dianalytic structure such that the map ker of W onto V is a morphism of Klein surfaces. [For a definition, see p. 17 of Alling and Greenleaf (1971).] Theorem 1.8.4 of Alling and Greenleaf (1971) is proved for a group G of automorphisms on a Klein surface that act discontinuously on it. In the present case, W can be regarded as a Klein surface, and $G = \{i_0, \tau_0\}$, where i_0 denotes the identity map on W. Since ker is a morphism, V is finite and connected. It also follows that $\operatorname{Re} \hat{f}$, $f \in A$, is a bounded harmonic function on V. By Proposition 1.9.1 of Alling and Greenleaf (1971), W is canonically isomorphic to the complex double V_c of V. Scanning carefully through the proof of Theorem 1.8.4 of Alling and Greenleaf (1971), it can be seen that only fixed points of τ_0 are sent to the boundary of V by ker (as W has no boundary points). Hence if τ_0 does not have any fixed point, then the boundary of V is empty. That V is nonorientable follows from Lemma 1.6.3 of Alling and Greenleaf (1971). In this case, the orienting double V_0 of V is the same as the complex double V_c of V, which is isomorphic to W.

$\square$

Finally, we give two examples to show that if the involution τ_0 does have fixed points, then the Klein surface V can be either orientable or nonorientable.

Example 3.5.4 Let A be the real disk algebra of Example 1.3.10. Let ϕ be the evaluation map at 0. Then $W := Q_A(\phi)$ is the open unit disk and $\tau_0 \colon W \to W$ is given by $\tau_0(z) = \bar{z}$. The set F of fixed points of τ_0 is the open interval $(-1,1)$ and $P_A(\ker(\phi)) = V = W/\tau_0$ is

$$\{z \colon |z| < 1, \quad \text{Im } z \geq 0\},$$

which is orientable.

Example 3.5.5 Let $w = 1 + i$ and $L = \{n + mw \colon n,m \text{ integers}\}$. Since $\bar{w} = 1 - i = 2 - w \in L$, the map $z \to \bar{z}$ descends to the complex torus $S = \mathbb{C}/L$. Let $\tau \colon S \to S$ be the map induced by $z \to -i\bar{z}$. Then τ is antianalytic. The set F of fixed points of τ is given by $F = \{tw \colon t \in \mathbb{R}\}$. Hence F is a circle in S. Let D be an open disk in S such that $\bar{D} \cap F$ is empty. Then $X = S \backslash (D \cup \tau(D))$ is a compact Riemann surface with boundary and $\tau_{|S}$ is an antianalytic involution on X. Let A be the real function algebra on (X,τ) as described in Example 1.3.12. Let ϕ be the evaluation map at an interior point of X. Then $W := Q_A(\phi) = X^0$ and $\tau_0 = \tau_{|X^0}$. In this case $P_A(\ker(\phi)) = V = W/\tau_0$ is a Möbius strip with one disk removed, which is nonorientable [see Example 1.6.3, Proposition 1.9.1, and Corollary 1.9.3 of Alling and Greenleaf (1971)].

Results in this section were proved by Kulkarni and Limaye (1981b).

Remark 3.5.6 In the case of a complex function algebra B, it was proved by Wermer that if Re B is dense in $C_{\mathbb{R}}(X)$ and if P is a nontrivial Gleason part of B, then there is an analytic map of the open unit disk U onto P (Browder (1969, p. 244). In analogy with this, if Re A is dense in $C_{\mathbb{R}}(X,\tau) := \{u \in C_{\mathbb{R}}(X) \colon u \circ \tau = u\}$ and if Q is a nontrival Gleason part A, then we may expect to find an analytic/harmonic map onto Q.

Now let $B = \{f + ig \colon f,g \in A\}$. Since Re $B = \text{Re } A + \text{Im } A$ and $C_{\mathbb{R}}(X) = C_{\mathbb{R}}(X,\tau) + C_S(X,\tau)$, where $C_S(X,\tau) := \{v \in C_{\mathbb{R}}(X) \colon v \circ \tau = -v\}$, it is easy to see that the following statements are equivalent:
(i) Re B is dense in $C_{\mathbb{R}}(X)$.
(ii) Re A is dense in $C_{\mathbb{R}}(X,\tau)$ and Im A is dense in $C_S(X,\tau)$.

Hwang (1990) has given a detailed proof of the equivalence of (i) and (ii) and asked whether the denseness of Re A in $C_{\mathbb{R}}(X,\tau)$ implies the denseness of Im A in $C_S(X,\tau)$, and vice versa. If it does, then any one of these imply that Re B is dense in $C_{\mathbb{R}}(X)$ and we can use the method of proof of Theorem 3.5.1 (the complexification technique combined with Wermer's theorem) to deduce the existence of an analytic map onto a nontrival Gleason part Q of A.

Kulkarni and Limaye (1981b) mention the problem of finding weaker conditions on A that imply the existence of a harmonic map into Car(A) or into $M(A)$, without necessarily implying the existence of an analytic map into Car(A).

4

Boundaries for a Real Function Algebra

In Chapter 3 we discussed the problem of embedding an analytic structure in the carrier space of a real function algebra. Even when such an analytic structure cannot be embedded, we can still ask whether the (Gelfand transforms of) functions in a real function algebra exhibit some properties of analytic functions. By the maximum modulus principle, we know that the functions in the real disk algebra attain their maximum absolute value on the unit circle. The aim of the present chapter is to study subsets on which every function attains its maximum absolute value. Such a subset is called a boundary. If a boundary is closed, we can regard the given algebra as a function algebra on this boundary. Hence we look for a minimal closed boundary. Our search for a minimal closed boundary proceeds via the study of subsets on which the real parts of functions assume their maximum value. Such a subset is called a Choquet set.

In the first section we consider a real subspace A of $C(X,\tau)$ containing the constant function 1 and the associated subset $\mathrm{Ch}(A)$ of X consisting of all x in X such that Re e_x has a unique RPR measure. Several characterizations of $\mathrm{Ch}(A)$ are given. They are employed to show that $\mathrm{Ch}(A)$ is contained in every closed Choquet set for A and if Re A separates the points of X/τ, then $\mathrm{Ch}(A)$ itself is a Choquet set for A. Its closure, then, is the smallest closed Choquet set for A. This development is followed by an application of these characterizations to derive a classical theorem of Choquet.

136

The second section begins with some relationships between a boundary for A and a Choquet set for A. In particular, it is shown that if A is a uniformly closed real subalgebra of $C(X,\tau)$ containing 1, then every Choquet set for A is a boundary for A, and vice versa. If Re A separates the points of X/τ, then Ch(A) is a boundary for A, called the Choquet boundary, and its closure is the smallest closed boundary for A, called the Shilov boundary.

In the third section we indicate how the well-known theory for a complex subspace (or subalgebra) B of $C(X)$ is developed. Then we study the relationship between Ch(A) and Ch(B), where A is a subspace of $C(X,\tau)$ containing 1 and $B := \{f + ig : f, g \in A\}$. A result of particular interest here is that Ch(A) and Ch(B) coincide when A is a real function algebra on (X,τ). We also compute Choquet and Shilov boundaries of some real and complex algebras.

4.1 CHOQUET SETS FOR SUBSPACES OF $C(X,\tau)$

Throughout this and the next section, X is a compact Hausdorff space, τ is a topological involution on X, and A is a (real) subspace of $C(X,\tau)$ containing 1. A subset S of X is called τ-*invariant* if $\tau(S) = S$.

Definition 4.1.1 A subset S of X is called a *Choquet set for A* if it is τ-invariant and if Re f assumes its maximum on S for every f in A.

Clearly, X itself is a Choquet set for A. In the search for smaller Choquet sets for A, the following subset of X turns out to be of significance.

If $x \in X$, then the real part of the evaluation map on A at x, Re e_x, belongs to

$$K(A) := \{\psi \in A' : \psi(1) = 1 = \|\psi\|\}.$$

Recalling Definition 3.2.6 of a real part representing (RPR) measure for an element of $K(A)$ and Theorem 3.2.7(a), we let

$$Ch(A) := \{x \in X : Re\ e_x \text{ has a unique RPR measure on } X\}.$$

Note that for every $x \in X$, the measure

$$m_x := \frac{\delta_x + \delta_{\tau(x)}}{2}$$

is a RPR measure for Re e_x. If $x \in Ch(A)$, then this is the only RPR measure for Re e_x. Observe that if $x = \tau(x)$, then $m_x = \delta_x$. Also, if $x \in Ch(A)$, then $\tau(x) \in Ch(A)$. Thus Ch(A) is τ-invariant.

In this section our aim is to show that (i) Ch(A) is contained in every closed Choquet set for A, and that (ii) Ch(A) is itself a Choquet set for A, provided that Re A separates the points of X/τ. (See Definition 1.3.6.) For this purpose we first take up a study of RPR measures. Since every RPR measure μ satisfies $\mu \circ \tau = \mu$, the following result is useful while considering the uniqueness of RPR measures.

Lemma 4.1.2 Let μ_1 and μ_2 be measures on X such that $\mu_1 \circ \tau = \mu_1$, $\mu_2 \circ \tau = \mu_2$, and

$$\int_X u \, d\mu_1 = \int_X u \, d\mu_2$$

for all $u \in C_\mathbb{R}(X)$ satisfying $u \circ \tau = u$. Then $\mu_1 = \mu_2$.

Proof. By the uniqueness part of the Riesz representation theorem for $C(X)$, it is enough to show that $\int_X h \, d\mu_1 = \int_X h \, d\mu_2$ for all $h \in C(X)$. Considering the real and imaginary parts of a function in $C(X)$, it is sufficient to prove that $\int_X w \, d\mu_1 = \int_X w \, d\mu_2$ if $w \in C_\mathbb{R}(X)$. But $w = u + v$, where $u = (w + w \circ \tau)/2$ and $v = (w - w \circ \tau)/2$. Since $u \circ \tau = u$, we have $\int_X u \, d\mu_1 = \int_X u \, d\mu_2$. Since $v \circ \tau = -v$ and $\mu_1 \circ \tau = \mu_1$, we see that

$$\int_X v \, d\mu_1 = -\int_X v \circ \tau \, d\mu_1 = -\int_X v \, d(\mu_1 \circ \tau) = -\int_X v \, d\mu_1;$$

that is, $\int_X v \, d\mu_1 = 0$. Similarly, $\int_X v \, d\mu_2 = 0$. Thus $\int_X w \, d\mu_1 = \int_X w \, d\mu_2$. $\square$

Theorem 4.1.3 If Re A is uniformly dense in $\{u \in C_\mathbb{R}(X) : u \circ \tau = u\}$, then Ch($A$) = X. In particular, Ch($C(X,\tau)$) = X.

Proof. Let $x \in X$ and μ_1, μ_2 be RPR measures for Re e_x. Then for all f in A,

$$\int_X \text{Re} \, f \, d\mu_1 = \text{Re} \, f(x) = \int_X \text{Re} \, f \, d\mu_2.$$

Let $u \in C_\mathbb{R}(X)$ with $u \circ \tau = u$. Then there is a sequence (f_n) in A such that $\|\text{Re} \, f_n - u\|_\infty \to 0$, so that $\int_X \text{Re} \, f_n \, d\mu_j \to \int_X u \, d\mu_j$ for $j = 1,2$. Thus $\int_X u \, d\mu_1 = \int_X u \, d\mu_2$. Now, the result follows from Lemma 4.1.2. $\square$

Here is an example of $x \in X \backslash \text{Ch}(A)$. Let X be the closed unit disk and A be the real disk algebra of Example 1.3.10. Let $x = 0$. By the Cauchy integral formula,

$$f(0) = \frac{1}{2\pi i} \int_{|z|=1} \frac{f(z)}{z} \, dz = \frac{1}{2\pi} \int_0^{2\pi} f(e^{i\theta}) \, d\theta$$

for every f in A, so that

$$\operatorname{Re} e_0(f) = \operatorname{Re} f(0) = \int_0^{2\pi} \operatorname{Re} f(e^{i\theta}) \frac{d\theta}{2\pi}.$$

Let $\mu = d\theta/2\pi$, supported on the unit circle. Then μ is a probability measure on X, $\mu \circ \tau = \mu$, where $\tau(z) = \bar{z}$, and μ is, in fact, a RPR measure for $\operatorname{Re} e_0$, which is different from $m_0 = \delta_0$.

Let $\psi \in K(A)$, and $u \in C_{\mathbb{R}}(X)$ with $u \circ \tau = u$. We let

$$\alpha_u(\psi) := \sup\{\psi(f) : f \in A, \quad \operatorname{Re} f \leq u\}$$

and

$$\beta_u(\psi) := \inf\{\psi(f) : f \in A, \quad \operatorname{Re} f \geq u\}.$$

If $f, g \in A$ with $\operatorname{Re} f \leq u$ and $\operatorname{Re} g \geq u$, then $\operatorname{Re}(g - f) \geq 0$, and by Lemma 3.2.4, $\psi(g - f) \geq 0$, that is, $\psi(g) \geq \psi(f)$. This shows that $\alpha_u(\psi) \leq \beta_u(\psi)$.

Lemma 4.1.4 Let $\psi \in K(A)$.
(a) If $u \in C_{\mathbb{R}}(X)$ with $u \circ \tau = u$, then there exists a RPR measure μ for ψ with $\int_X u \, d\mu = \gamma$ if and only if $\alpha_u(\psi) \leq \gamma \leq \beta_u(\psi)$.
(b) ψ has a unique RPR measure if and only if $\alpha_u(\psi) = \beta_u(\psi)$ for every $u \in C_{\mathbb{R}}(X)$ with $u \circ \tau = u$, and in that case

$$\alpha_u(\psi) = \int_X u \, d\mu = \beta_u(\psi),$$

where μ is the unique RPR measure for ψ.

Proof.
(a) Let $u \in C_{\mathbb{R}}(X)$ with $u \circ \tau = u$. If μ is a RPR measure for ψ, and $f \in A$ with $\operatorname{Re} f \leq u$, then

$$\psi(f) = \int_X \operatorname{Re} f \, d\mu \leq \int_X u \, d\mu,$$

so that $\alpha_u(\psi) \leq \int_X u \, d\mu$. Similarly, $\int_X u \, d\mu \leq \beta_u(\psi)$. This proves the "only if" part.

Conversely, let γ be such that $\alpha_u(\psi) \leq \gamma \leq \beta_u(\psi)$. First consider the case when $u \in A$. Then it is easy to see that $\alpha_u(\psi) = \psi(u) = \beta_u(\psi)$, so that $\gamma = \psi(u)$. By Theorem 3.2.7, there exists a RPR measure μ for ψ, and then

$$\gamma = \psi(u) = \int_X \mathrm{Re}\, u \, d\mu = \int_X u \, d\mu.$$

Next, consider the case when $u \notin A$. Let

$$A_u := \{f + tu : f \in A, \quad t \in \mathbb{R}\}$$

be the (real) subspace of $C(X,\tau)$ spanned by A and u. Define $\theta : A_u \to \mathbb{R}$ by

$$\theta(f + tu) = \psi(f) + t\gamma.$$

Then θ is well-defined, linear, and $\theta(1) = 1$. Let $f \in A$ and $t \in \mathbb{R}$ be such that $\mathrm{Re}\, f + tu \geq 0$. We show that $\psi(f) + t\gamma \geq 0$. If $t = 0$, this follows from Lemma 3.2.4 as $\psi \in K(A)$. If $t > 0$, then $u \geq -\mathrm{Re}\, f/t = \mathrm{Re}(-f/t)$, so that $\psi(-f/t) \leq \alpha_u(\psi) \leq \gamma$; that is, $0 \leq \psi(f) + t\gamma$, as desired. If $t < 0$, then $u \leq \mathrm{Re}(-f/t)$, so that $\gamma \leq \beta_u(\psi) \leq \psi(-f/t)$; that is, $\psi(f) + t\gamma \geq 0$, as desired. Thus, whenever $f + tu \in A_u$ and $\mathrm{Re}(f + tu) \geq 0$, we see that $\mathrm{Re}\,\theta(f + tu) = \theta(f + tu) = \psi(f) + t\gamma \geq 0$. Hence, by Lemma 3.2.4, $\theta \in K(A_u)$, and by Theorem 3.2.7, there is a RPR measure μ for θ. Since $\theta_{|A} = \psi$, and μ is a probability measure [Theorem 3.2.5(b)], it follows that μ is a RPR measure for ψ. Also,

$$\int_X u \, d\mu = \int_X \mathrm{Re}\, u \, d\mu = \theta(u) = \gamma.$$

This proves the "if" part.

(b) If for some $u \in C_\mathbb{R}(X)$ with $u \circ \tau = u$, we have $\alpha_u(\psi) < \beta_u(\psi)$, then we can find real numbers γ_1 and γ_2 such that $\alpha_u(\psi) < \gamma_1 < \gamma_2 < \beta_u(\psi)$. By (a), there are RPR measures μ_1 and μ_2 for ψ with $\int_X u \, d\mu_1 = \gamma_1$ and $\int_X u \, d\mu_2 = \gamma_2$. Then, clearly, $\mu_1 \neq \mu_2$.

Conversely, assume that $\alpha_u(\psi) = \beta_u(\psi)$ for all $u \in C_\mathbb{R}(X)$ with $u \circ \tau = u$. If μ is a RPR measure for ψ, then for every u in $C_\mathbb{R}(X)$ with $u \circ \tau = u$ and every f in A with $\mathrm{Re}\, f \leq u$, we have

$$\psi(f) = \int_X \mathrm{Re}\, f \, d\mu \leq \int_X u \, d\mu,$$

so that $\alpha_u(\psi) \leq \int_X u \, d\mu$, and similarly, $\int_X u \, d\mu \leq \beta_u(\psi)$. But since $\alpha_u(\psi) = \beta_u(\psi)$, it follows that $\int_X u \, d\mu = \alpha_u(\psi)$ for all $u \in C_\mathbb{R}(X)$ with $u \circ \tau = u$. By Lemma 4.1.2, there is a unique RPR measure for ψ. $\qquad\square$

Theorem 4.1.5 [Compare Theorem 2.2.6 of Browder (1969).] Let $x \in X$. Then the following statements are equivalent:

(i) $x \in \mathrm{Ch}(A)$.

(ii) $\alpha_u(\mathrm{Re}\ e_x) = u(x) = \beta_u(\mathrm{Re}\ e_x)$ for every $u \in C_{\mathbb{R}}(X)$ with $u \circ \tau = u$.

(iii) For every α and β with $0 < \alpha < \beta$ and for every τ-invariant neighborhood U of $\{x,\tau(x)\}$, there exists f in A with $\mathrm{Re}\ f \leq 0$, $\mathrm{Re}\ f(x) > -\alpha$ and $\mathrm{Re}\ f(y) < -\beta$ for all y in $X \backslash U$.

(iv) There exist α and β with $0 < \alpha < \beta$ such that for every τ-invariant neighborhood U of $\{x,\tau(x)\}$, there exists f in A with $\mathrm{Re}\ f \leq 0$, $\mathrm{Re}\ f(x) > -\alpha$ and $\mathrm{Re}\ f(y) < -\beta$ for all y in $X \backslash U$.

(v) If μ is a RPR measure for $\mathrm{Re}\ e_x$, then $\mu(\{x\}) > 0$.

Proof. Statements (i) and (ii) are equivalent by Lemma 4.1.4(b) with $\psi = \mathrm{Re}\ e_x$.

(ii) implies (iii): Let $0 < \alpha < \beta$ and U be a τ-invariant neighborhood of $\{x,\tau(x\}$. By Urysohn's lemma, there exists $w \in C_{\mathbb{R}}(X)$ such that $w \leq 0$, $w(x) = 0$ and $w(y) < -\sqrt{\beta}$ for all y in $X \backslash U$. For $s \in X$, let

$$u(s) := -w(s)w(\tau(s)).$$

Then $u \in C_{\mathbb{R}}(X)$, $u \circ \tau = u$, $u \leq 0$, $u(x) = 0$, and $u(y) < -\beta$ for all y in $X \backslash U$. Since

$$\alpha_u(\mathrm{Re}\ e_x) = \sup\{\mathrm{Re}\ f(x): f \in A, \mathrm{Re}\ f \leq u\}$$
$$= u(x) = 0,$$

we see that there exists f in A with $\mathrm{Re}\ f \leq u \leq 0$, $\mathrm{Re}\ f(x) \geq -\alpha$ and $\mathrm{Re}\ f(y) < -\beta$ for all y in $X \backslash U$.

(iii) implies (iv): Obvious.

(iv) implies (v): Let μ be a RPR measure for $\mathrm{Re}\ e_x$. If U is a τ-invariant neighborhood of $\{x,\tau(x)\}$, then there is some f in A with $\mathrm{Re}\ f \leq 0$, $\mathrm{Re}\ f(x) > -\alpha$, and $\mathrm{Re}\ f(y) < -\beta$ for all y in $X \backslash U$, so that

$$-\alpha < \mathrm{Re}\ f(x) = \int_X \mathrm{Re}\ f\, d\mu$$

$$= \int_U \mathrm{Re}\ f\, d\mu + \int_{X \backslash U} \mathrm{Re}\ f\, d\mu$$

$$\leq 0 - \beta\mu(X \backslash U) = -\beta(1 - \mu(U)).$$

Thus $\mu(U) \geq (\beta - \alpha)/\beta$ for every τ-invariant neighborhood U of $\{x,\tau(x)\}$. Hence by the outer regularity of μ,

$$\mu(\{x,\tau(x)\}) \geq \frac{\beta - \alpha}{\beta} > 0.$$

(v) implies (i): If $x \notin Ch(A)$, then there is a RPR measure v for Re e_x such that $v \neq m_x := (\delta_x + \delta_{\tau(x)})/2$. Let $c = v(\{x,\tau(x)\})$, so that $0 \le c < 1$. Let

$$\mu := \frac{v - cm_x}{1 - c}.$$

It can easily be verified that $\mu \circ \tau = \mu$ and that for all f in A,

$$\int_X \text{Re} f \, d\mu = \text{Re} f(x).$$

Also, v is a probability measure by Theorem 3.2.5(b), and

$$m_x(E) = 1, \qquad v(E) \ge c \qquad \text{if } x \text{ and } \tau(x) \text{ belong to } E,$$

$$m_x(E) = \tfrac{1}{2}, \qquad v(E) \ge \frac{c}{2} \qquad \text{if only one of } x, \tau(x) \text{ belongs to } E,$$

$$m_x(E) = 0, \qquad v(E) \ge 0 \qquad \text{if neither } x \text{ nor } \tau(x) \text{ belongs to } E.$$

It follows that $\mu(X) = 1$, and $\mu(E) \ge 0$ for all Borel subsets E of X. Hence μ is a RPR measure for Re e_x. Since $\mu(\{x,\tau(x)\}) = 0$, we are through.   □

Corollary 4.1.6   Ch(A) is contained in every closed Choquet set for A.

Proof.   Let S be a closed Choquet set for A and $x \in Ch(A)$. If $x \notin S$, then there is a neighborhood v of x such that $V \cap S = \phi$. Since $S = \tau(S)$, the τ-invariant neighborhood $U = V \cup \tau(V)$ of $\{x,\tau(x)\}$ satisfies $U \cap S = \phi$. By (iv) of Theorem 4.1.5, there exist α and β with $0 < \alpha < \beta$ and f in A such that Re $f(x) > -\alpha$ and Re $f(y) < -\beta$ for all y in $X \backslash U$, and in particular, for all y in S. But then Re f does not assume its maximum on S, a contradiction to S being a Choquet set for A. This shows that Ch(A) is contained in S.   □

Corollary 4.1.7   Let $x \in X$. Suppose that there exists $c < 1$ such that for every τ-invariant neighborhood U of $\{x,\tau(x)\}$, there is g in A such that $g(x) = 1 = \|g\|$ and $|g(y)| < c$ for all y in $X \backslash U$. Then $x \in Ch(A)$.

Proof.   Find d such that $c < d < 1$. The result follows by letting $f = g - 1$, $\alpha = 1 - d$, and $\beta = 1 - c$ in the statement (iv) of Theorem 4.1.5.   □

Corollary 4.1.8   If X is metrizable, then Ch(A) is a G_δ set in X.

Proof.   Let $d(\cdot,\cdot)$ be a metric on X that induces the topology of X. For x

$\in X$ and $n = 1,2,\ldots$, let

$$V_{x,n} := \left\{y \in X : d(x,y) < \frac{1}{n}\right\},$$

$$U_{x,n} := V_{x,n} \cup \tau(V_{x,n}),$$

and let G_n be the set of all x in X for which there exists f in A with $\operatorname{Re} f \leq 0$, $\operatorname{Re} f(x) > -1$, and $\operatorname{Re} f(y) < -2$ for all y in $X \backslash U_{x,n}$. Then each G_n is open, and $\operatorname{Ch}(A)$ is the intersection of all the G_n's by (iii) and (iv) of Theorem 4.1.5. $\qquad \square$

Remark 4.1.9 The set $\operatorname{Ch}(A)$ can be empty for some real subspaces of $C(X,\tau)$ containing 1, as the trivial example $A = \{$all real constants$\}$ shows. To guarantee that $\operatorname{Ch}(A)$ is not empty and is, in fact, a Choquet set for A, we give another characterization of $\operatorname{Ch}(A)$ under a separation condition on $\operatorname{Re} A$.

Recall from Definition 1.3.6 that a subset E of $\{u \in C_{\mathbb{R}}(X) : u \circ \tau = u\}$ is said to *separate x from other points* of X/τ if for y in X with $y \neq x$, $y \neq \tau(x)$, there is u in E such that $u(x) \neq u(y)$. E is said to *separate the points of X/τ* if E separates every $x \in X$ from other points of X/τ.

Theorem 4.1.10
(a) If $x \in \operatorname{Ch}(A)$, then $\operatorname{Re} e_x$ is an extreme point of $K(A)$.
(b) If ψ is an extreme point of $K(A)$, then $\psi = \operatorname{Re} e_x$ for some $x \in X$, and if $\operatorname{Re} A$ separates x from other points of X/τ, then, in fact, $x \in \operatorname{Ch}(A)$.

Proof.
(a) Let $x \in \operatorname{Ch}(A)$ and $\psi_1, \psi_2 \in K(A)$ be such that $\operatorname{Re} e_x = t\psi_1 + (1 - t)\psi_2$ with $0 < t < 1$. Let μ_1 and μ_2 be RPR measures for ψ_1 and ψ_2, respectively. Then $t\mu_1 + (1 - t)\mu_2$ is a RPR measure for $\operatorname{Re} e_x$. Since $x \in \operatorname{Ch}(A)$, we see that $t\mu_1 + (1 - t)\mu_2 = m_x$. If E is a Borel subset of X with $x \notin E$ and $\tau(x) \notin E$, then $0 = m_x(E) = t\mu_1(E) + (1 - t)\mu_2(E)$. As μ_1 and μ_2 are positive measures by Theorem 3.2.5(b), we have $\mu_1(E) = 0 = \mu_2(E)$. Also, $\mu_1 \circ \pi = \mu_1$ and $\mu_2 \circ \tau = \mu_2$. This shows that $\mu_1 = m_x = \mu_2$, and in turn, $\psi_1 = \operatorname{Re} e_x = \psi_2$. Thus $\operatorname{Re} e_x$ is an extreme point of $K(A)$.
(b) Let ψ be an extreme point of $K(A)$. Let μ be a RPR measure for ψ and suppose that $x \in \operatorname{Supp}(\mu)$. Then $\tau(x) \in \operatorname{Supp}(\mu)$. Note that μ is a probability measure and $\mu \circ \tau = \mu$. If $\mu(U) = 1$ for every τ-invariant neighborhood U of $\{x, \tau(x)\}$, then clearly $\mu = m_x$.
If for some τ-invariant neighborhood U of $\{x, \tau(x)\}$, we have $\mu(U) < 1$, then let

$$\theta_1(f) := \frac{1}{\mu(U)} \int_U \operatorname{Re} f \, d\mu \quad \text{and} \quad \theta_2 := \frac{1}{\mu(X \setminus U)} \int_{X \setminus U} \operatorname{Re} f \, d\mu$$

for f in A. Then θ_1 and θ_2 are in $K(A)$ and

$$\psi = \mu(U)\theta_1 + (1 - \mu(U))\theta_2.$$

Since ψ is an extreme point of $K(A)$, we have $\theta_1 = \psi = \theta_2$. Thus $\psi(f) = (1/\mu(U)) \int_U \operatorname{Re} f \, d\mu$ for all f in A.

Since $\mu(U) < 1$, we have $\mu(V) < 1$ for any smaller τ-invariant neighborhood V of $\{x, \tau(x)\}$. Then, as above, we can prove that

$$\psi(f) = \frac{1}{\mu(V)} \int_V \operatorname{Re} f \, d\mu \tag{*}$$

for all f in A.

Now, let $\varepsilon > 0$ and

$$V_\varepsilon := \{y \in X : \operatorname{Re} f(x) - \varepsilon < \operatorname{Re} f(y) < \operatorname{Re} f(x) + \varepsilon\}.$$

Then $V_\varepsilon = \tau(V_\varepsilon)$ and V_ε is a neighborhood of $\{x, \tau(x)\}$. Since $x \in \operatorname{Supp}(\mu)$, we see that $\mu(V_\varepsilon) > 0$ and

$$(\operatorname{Re} f(x) - \varepsilon)\mu(V_\varepsilon) < \int_{V_\varepsilon} \operatorname{Re} f \, d\mu < (\operatorname{Re} f(x) + \varepsilon)\mu(V_\varepsilon),$$

so that

$$\operatorname{Re} f(x) - \varepsilon < \frac{1}{\mu(V_\varepsilon)} \int_{V_\varepsilon} \operatorname{Re} f \, d\mu < \operatorname{Re} f(x) + \varepsilon;$$

that is, $\operatorname{Re} f(x) - \varepsilon < \psi(f) < \operatorname{Re} f(x) + \varepsilon$ by (*). Since this is true for all $\varepsilon > 0$, we have $\operatorname{Re} f(x) = \psi(f)$; that is, $\psi = \operatorname{Re} e_x$.

Finally, suppose that $\operatorname{Re} A$ separates x from other points of X/τ. Then $\operatorname{Supp}(\mu) = \{x, \tau(x)\}$. For if $y \in \operatorname{Supp}(\mu)$, then we can prove, as above, that $\psi = \operatorname{Re} e_y$. But then $\operatorname{Re} e_y = \operatorname{Re} e_x$ implies that $y = x$ or $\tau(x)$. This shows that $\mu = m_x$; that is, $x \in \operatorname{Ch}(A)$. $\square$

Corollary 4.1.11 Assume that $\operatorname{Re} A$ separates the points of X/τ. Then the following statements hold.
(a) $\operatorname{Ch}(A) = \{x \in X : \operatorname{Re} e_x \text{ is an extreme point of } K(A)\}$ and is nonempty.
(b) $\operatorname{Ch}(A)$ is a Choquet set for A.
(c) The closure of $\operatorname{Ch}(A)$ is the smallest closed Choquet set for A.

Proof.
(a) This follows from Theorem 4.1.10 by the Krein–Milman theorem

for the nonempty weak*-closed convex subset $K(A)$ of the unit ball of A'.

(b) Let $f \in A$. Since $\operatorname{Re} f$ is continuous on the compact space X, there is x in X such that $\operatorname{Re} f(x) = \max \{\operatorname{Re} f(y) : y \in X\}$. Let

$$L := \{\psi \in K(A) : \psi(f) = \operatorname{Re} f(x)\}.$$

Then L is a weak*-closed convex subset of the unit ball of A' and it is nonempty since $\operatorname{Re} e_x \in L$. By the Krein–Milman theorem, there is an extreme point ψ of L. We show that ψ is, in fact, an extreme point of $K(A)$. Let $\psi_1, \psi_2 \in K(A)$ and $\psi = t\psi_1 + (1 - t)\psi_2$ for some t, $0 < t < 1$. Since $\psi \in L$, we have

$$\operatorname{Re} f(x) = \psi(f) = t\psi_1(f) + (1 - t)\psi_2(f). \tag{*}$$

Since $\operatorname{Re}(\operatorname{Re} f(x) - f) = \operatorname{Re} f(x) - \operatorname{Re} f \geq 0$ and $\psi_j \in K(A)$, it follows from Lemma 3.2.4 that $\psi_j(\operatorname{Re} f(x) - f) = \operatorname{Re} \psi_j(\operatorname{Re} f(x) - f) \geq 0$; that is, $\operatorname{Re} f(x) \geq \psi_j(f)$ for $j = 1,2$. Now (*) shows that $\operatorname{Re} f(x) = \psi_j(f)$, so that $\psi_j \in L$ for $j = 1,2$. But as ψ is an extreme point of L, we have $\psi_1 = \psi = \psi_2$; that is, ψ is an extreme point of $K(A)$. By part (a), $\psi = \operatorname{Re} e_y$ for some $y \in \operatorname{Ch}(A)$. Since

$$\operatorname{Re} f(y) = \psi(f) = \operatorname{Re} f(x) = \max\{\operatorname{Re} f(s) : s \in X\},$$

we see that $\operatorname{Re} f$ assumes its maximum at $y \in \operatorname{Ch}(A)$. $\operatorname{Ch}(A)$ is thus a Choquet set for A.

(c) Since $\operatorname{Ch}(A)$ is a Choquet set for A, so is the closure of $\operatorname{Ch}(A)$. Also, $\operatorname{Ch}(A)$ and hence its closure are contained in every closed Choquet set for A by Corollary 4.1.6. This shows that the closure of $\operatorname{Ch}(A)$ is the smallest closed Choquet set for A. $\qquad\square$

Remark 4.1.12 Let A_0 be a subspace of $C(X, \tau)$. (A_0 may not contain 1.) The method of proof of Theorem 4.1.10 can be employed to investigate the extreme points of the closed unit ball of A_0'. Let

$$L(A_0) := \{\psi \in A_0' := \|\psi\| \leq 1\},$$
$$F := \{x \in X : \tau(x) = x\},$$
$$\Lambda := \{\lambda \in \mathbb{C} : |\lambda| = 1, \operatorname{Im} \lambda \geq 0\},$$
$$\check{X} := (F \times \{-1,1\}) \cup ((X\backslash F) \times \Lambda).$$

For (x, λ) in $\check{X}$, define

$$\psi_{x,\lambda}(f) := \operatorname{Re}(\lambda f(x)), \quad f \in A_0,$$
$$m_{x,\lambda} := \frac{\lambda \delta_x + \bar{\lambda} \delta_{\tau(x)}}{2}.$$

Then $\psi_{x,\lambda} \in L(A_0)$ and for every f in A_0,

$$\psi_{x,\lambda}(f) = \int_X f \, dm_{x,\lambda}$$

and

$$m^*_{x,\lambda} = m_{x,\lambda}.$$

The following results are proved in Kulkarni and Limaye (1991b).

(a) If $m_{x,\lambda}$ is the only representing measure μ for $\psi_{x,\lambda}$ satisfying $\mu^* = \mu$, then $\psi_{x,\lambda}$ is an extreme point of $L(A_0)$.

(b) If ψ is an extreme point of $L(A_0)$, then $\psi = \psi_{x,\lambda}$ for some x,λ in $\tilde{X}$, and if for every (y,λ) in $\tilde{X}$, with $(y,\lambda') \neq (x,\lambda)$, there exists f in A_0 such that

$$\text{Re}(\lambda \, f(x)) \neq \text{Re}(\lambda' \, f(y)),$$

then $m_{x,\lambda}$ is the only representing measure μ for ψ satisfying $\mu^* = \mu$.

If $A_0 = C(X,\tau)$, then for every $(x,\lambda) \in \tilde{X}$, $m_{x,\lambda}$ is the only representing measure for $\psi_{x,\lambda}$ and the separation condition stated above is also satisfied. Hence ψ is an extreme point of $L(C(X,\tau))$ if and only if $\psi = \psi_{x,\lambda}$ for some $(x,\lambda) \in \tilde{X}$. This result is proved by Grzesiak (1986).

We conclude this section by considering the main theorem about the set $\text{Ch}(A)$: If X is metrizable, then for each $\psi \in K(A)$, there exists a RPR measure μ for ψ concentrated on $\text{Ch}(A)$. This result follows from a version of Choquet's theorem.

Lemma 4.1.13 Let $\psi \in K(A)$ and $u \in C_{\mathbb{R}}(X)$ with $u \circ \tau = u$. For $x \in X$, let

$$u_0(x) := \sup\{\text{Re} \, f(x) : f \in A, \, \text{Re} \, f \leq u\},$$

and let $E_u := \{x \in X : u(x) = u_0(x)\}$. Then ψ has a RPR measure μ such that $\mu(E_u) = 1$.

Proof. By Lemma 4.1.4(a), there is a RPR measure μ for ψ such that

$$\int_X u \, d\mu = \alpha_u(\psi).$$

Recalling that $\alpha_u(\psi) = \sup\{\psi(f) : f \in A, \, \text{Re} \, f \leq u\}$, we find that for $\varepsilon > 0$, there is $f \in A$ such that $\text{Re} \, f \leq u$ and $\alpha_u(\psi) - \varepsilon < \psi(f)$. Hence $\text{Re} \, f \leq u_0$, and we have

$$\int_X u \, d\mu - \varepsilon = \alpha_u(\psi) - \varepsilon < \psi(f) = \int_X \text{Re} f \, d\mu \le \int_X u_0 \, d\mu.$$

Since this is true for every $\varepsilon > 0$, we have

$$\int_X u \, d\mu \le \int_X u_0 \, d\mu.$$

But since $u_0 \le u$ by the definition of u_0, we see that

$$0 = \int_X (u - u_0) \, d\mu = \int_{X \setminus E_u} (u - u_0) \, d\mu,$$

so $\mu(X \setminus E_u) = 0$; that is, $\mu(E_u) = 1$. $\qquad\square$

Lemma 4.1.14 Let X be a compact Hausdorff space.
(a) X is metrizable if and only if $C(X)$ is separable.
(b) If X is metrizable, A is a real subspace of $C(X)$ containing 1 and $K(A) = \{\psi \in A' : \|\psi\| = 1 = \psi(1)\}$, then $K(A)$ is compact and metrizable in the weak* topology.

Proof.
(a) This is well known; see, for example, Theorem 5.1 of Burckel (1972).
(b) That $K(A)$ is compact in the weak* topology follows from the Banach–Alaoglu theorem. To show that $K(A)$ is metrizable, we proceed as follows. Since X is a compact metric space, $C(X)$ is separable by (a) and so is A. Let $\{f_n\}_{n=1}^{\infty}$ be a countable dense subset of A. For ψ_1 and ψ_2 in $K(A)$, let

$$d(\psi_1, \psi_2) := \sum_{n=1}^{\infty} \frac{1}{2^n} \frac{|\psi_1(f_n) - \psi_2(f_n)|}{1 + \|f_n\|}.$$

It can be checked that d is a metric on $K(A)$. Let $\psi_0 \in K(A)$ and $\varepsilon > 0$. To show that $\{\psi \in K(A) : d(\psi, \psi_0) < \varepsilon\}$ is open in the weak* topology, consider $\psi_1 \in K(A)$ with $d(\psi_1, \psi_0) < \varepsilon$. Find $\varepsilon_0 > 0$ such that $d(\psi, \psi_0) + \varepsilon_0 < \varepsilon$, and n_0 such that $\sum_{n=n_0+1}^{\infty} (2/2^n) < \varepsilon_0/2$. Then

$$\left\{ \psi \in K(A) : |\psi(f_n) - \psi_1(f_n)| < \frac{\varepsilon_0(1 + \|f_n\|)}{2} \text{ for } n = 1, \ldots, n_0 \right\}$$

is a weak* open neighborhood of ψ_1 and it is contained in $\{\psi \in K(A) : d(\psi, \psi_0) < \varepsilon\}$. Thus every open set in the metric topology on $K(A)$ is open in the weak* topology on $K(A)$. Since the weak* topology is compact and the metric topology is Hausdorff, the two topologies coincide. □

Theorem 4.1.15 (Choquet's theorem) (Phelps, 1966, p. 19) Let X_0 be a compact convex metrizable subset of a locally convex topological vector space V. Then the set E of extreme points of X_0 is a G_δ-set. For every $x_0 \in X_0$, there is a probability measure μ_0 on X_0 such that $\mu_0(E) = 1$, and

$$a(x_0) = \int_{X_0} a \, d\mu_0$$

for every continuous affine function a on X_0.

Proof. Let the induced topology on X_0 be the same as the topology given by a metric d, and let for $n = 1, 2, \ldots$

$$X_n := \left\{ x \in V : x = \frac{y + z}{2}, \ y \text{ and } z \text{ in } X_0 \text{ and } d(y,z) \geq \frac{1}{n} \right\}.$$

Then X_n is closed in V and a point x in X_0 is not an extreme point of X_0 if and only if $x \in X_n$ for some $n = 1, 2, \ldots$; that is, $V \backslash E = \bigcup_{n=1}^{\infty} X_n$. Thus $E = \bigcap_{n=1}^{\infty} (V \backslash X_n)$ is a G_δ-set.

Next, let $A_0 := \{ a \in C_{\mathbb{R}}(X_0) : a \text{ affine on } X_0 \}$. Note that "$a \in A_0$" means that "$a$ is a real-valued continuous function on X_0 such that

$$a(tx + (1 - t)y) = ta(x) + (1 - t)a(y)$$

for all $x, y \in X_0$ and $0 \leq t \leq 1$."

It is clear that A_0 is a subspace of $C_{\mathbb{R}}(X_0)$ and $1 \in A_0$. Also, if $x, y \in X_0$ and $x \neq y$, then by the Hahn–Banach theorem for the locally convex space V, there is $v' \in V'$ such that $v'(x) \neq v'(y)$. Since $a = v'|_{X_0}$ belongs to A_0 and $a(x) \neq a(y)$, we see that A_0 separates the points of X_0.

Since X_0 is metrizable, Lemma 4.1.14(a) shows that $C_{\mathbb{R}}(X_0)$ is separable and so is $\{ a \in A_0 : \|a\| = 1 \}$; let $\{a_n\}$ be a countable dense subset of it. Define $u = -\sum_{n=1}^{\infty} a_n^2 / n^2$. Then $u \in C_{\mathbb{R}}(X_0)$. Also, u is strictly concave on X_0; that is, if $y, z \in X_0$, $y \neq z$, and $0 < t < 1$, then

$$u(ty + (1 - t)z) > tu(y) + (1 - t)u(z).$$

This follows by noting that $a_n(y) \neq a_n(z)$ for some n, so that a_n is a nonconstant affine function on the segment $[y,z]$ and hence $-a_n^2$ is strictly concave on $[y,z]$.

Let $x_0 \in X_0$. Applying Lemma 4.1.13 with $X = X_0$, $\tau =$ the identity map, $A = A_0$, $\psi(a) = a(x_0)$, $a \in A_0$, and the function u defined above, we obtain a RPR measure μ_0 for ψ such that $\mu_0(E_u) = 1$, where $E_u := \{ x \in$

$X_0 : u_0(x) = u(x)\}$, $u_0(x) = \sup\{a(x) : a \in A_0, \text{ Re } a \leq u\}$ for $x \in X_0$. Now μ_0 is a probability measure and

$$a(x_0) = \int_X \text{Re } a \, d\mu_0 = \int_X a \, d\mu_0$$

for every $a \in A_0$. To show $\mu_0(E) = 1$, we prove that $E_u \subset E$. Let $x \in E_u$ and $x = ty + (1 - t)z$ with $y, z \in X_0$ and $0 < t < 1$. If $y \neq z$, then by the strict concavity of u,

$$\begin{aligned}
u(x) &= u(ty + (1 - t)z) \\
&> tu(y) + (1 - t)u(z) \\
&\geq tu_0(y) + (1 - t)u_0(z) \\
&\geq \sup\{ta(y) + (1 - t)a(z) : a \in A_0, \quad \text{Re } a \leq u\} \\
&= \sup\{a(ty + (1 - t)z) : a \in A_0, \quad \text{Re } a \leq u\} \\
&= u_0(x),
\end{aligned}$$

so that $x \notin E_u$, a contradiction. Thus $y = z$, showing that x is an extreme point of X_0; that is, $x \in E$. $\qquad\square$

Theorem 4.1.16 Let X be a compact metric space, τ be a topological involution on X, and A be a (real) subspace of $C(X,\tau)$ containing 1 such that Re A separates the points of X/τ. Then every $\psi_0 \in K(A)$ has a RPR measure ν_0 on X such that $\nu_0(\text{Ch}(A)) = 1$.

Proof. $K(A)$ is a subset of the dual A' of A. Since X is a compact metric space, $K(A)$ is compact and metrizable in the weak* topology by Lemma 4.1.14. Let $r : X \rightarrow K(A)$ be defined by $r(x) := \text{Re } e_x$, $x \in X$. If E denotes the set of all extreme points of $K(A)$, then $E = r(\text{Ch}(A))$ by Theorem 4.1.10, since Re A separates the points of X/τ.

Let $\psi_0 \in K(A)$. Applying Theorem 4.1.15 with $X_0 = K(A)$ and $x_0 = \psi_0$, we obtain a probability measure μ_0 on $K(A)$ such that $\mu_0(E) = 1$, and

$$a(\psi_0) = \int_{K(A)} a \, d\mu_0$$

for every continuous affine function a on $K(A)$. For a Borel subset S of X, define

$$\nu_0(S) := \mu_0(r(S)).$$

Then ν_0 is a probability measure on X since μ_0 is a probability measure. Also, $\nu_0(\text{Ch}(A)) = \mu_0(E) = 1$. Since $r \circ \tau = r$, we have $\nu_0 \circ \tau = \nu_0$. Now, for a fixed f in A, consider $a_f : K(A) \rightarrow \mathbb{R}$ defined by

$$a_f(\psi) := \psi(f), \qquad \psi \in K(A).$$

Then a_f is a continuous affine function on $K(A)$. Since $\mu_0(r(X)) \geq \mu_0(E) = 1$, we see that $\mu_0(K(A)) = 1 = \mu_0(r(X))$, and hence

$$\psi_0(f) = a_f(\psi_0) = \int_{K(A)} a_f \, d\mu_0 = \int_{r(X)} a_f \, d\mu_0$$

$$= \int_X a_f \circ r \, d(\mu_0 \circ r) = \int_X \operatorname{Re} f \, d\nu_0,$$

since $(a_f \circ r)(x) = a_f(\operatorname{Re} e_x) = \operatorname{Re} e_x(f) = \operatorname{Re} f(x)$ for all x in X. Thus ν_0 is a RPR measure for ψ_0 satisfying $\nu_0(\operatorname{Ch}(A)) = 1$. $\qquad \square$

4.2 BOUNDARIES FOR SUBSPACES OF $C(X,\tau)$

Definition 4.2.1 A subset S of X is called a *boundary* for A if it is τ-invariant and if $|f|$ assumes its maximum on S for every f in A.

If S is a closed boundary for A, then S is a compact Hausdorff space, $\tau_{|S}$ is a topological involution on S, and the map $f \rightarrow f_{|S}$ is an isometric linear bijection of A onto the subspace $A_{|S}$ of $C(S,\tau_{|S})$. It is therefore appropriate to look for a minimal closed boundary for A. Let

$$S(A) = \cap \{S : S \text{ is a closed boundary for } A\}.$$

The concept of a Choquet set and, in particular, the set $\operatorname{Ch}(A)$ introduced in Section 4.1 will now be used to study boundaries for A. In the special case where τ is the identity map [and hence $A \subset C_{\mathbb{R}}(X)$], the concept of a boundary for A coincides with that of a Choquet set for A because $|f|$ assumes its maximum at a point if and only if $\operatorname{Re} f = f$ or $\operatorname{Re}(-f) = -f$ assumes its maximum at that point.

Theorem 4.2.2
(a) If S is a boundary for A, then the closure of S is a Choquet set for A.
(b) The closure of $\operatorname{Ch}(A)$ is contained in $S(A)$.
(c) If $\operatorname{Re} A$ separates the points of X/τ, then $S(A)$ contains a Choquet set for A. In particular, $S(A)$ is nonempty.

Proof.
   (a) Let S be a boundary for A and $\bar{S}$ denote its closure in X. Then $\bar{S}$ is a boundary for A. Suppose that $\bar{S}$ is not a Choquet set for A. Then there is f in A such that $\operatorname{Re} f(x) < 1$ for all x in $\bar{S}$ and $\operatorname{Re} f(y) = 1$ for some y in $X \backslash \bar{S}$. Then

$$\delta := \sup\{\operatorname{Re} f(x) : x \in \bar{S}\} < 1.$$

Find $t > 0$ such that

$$t(\|f\|^2 - |f(y)|^2) < 2(1 - \delta),$$

and let $f_t := 1 + tf$. Then for every $x \in \bar{S}$,

$$\begin{aligned}
|f_t(x)|^2 &= 1 + 2t \operatorname{Re} f(x) + t^2|f(x)|^2 \\
&\leq 1 + 2t\delta + t^2\|f\|^2 \\
&< 1 + 2t + t^2|f(y)|^2 \\
&= 1 + 2t \operatorname{Re} f(y) + t^2|f(y)|^2 = |f_t(y)|^2,
\end{aligned}$$

since $\delta < 1$ and $\operatorname{Re} f(y) = 1$. This shows that $|f_t|$ does not assume its maximum on $\bar{S}$, a contradiction.

(b) By Corollary 4.1.6, $\operatorname{Ch}(A)$ is contained in every Choquet set for A, and hence, by (a), in every closed boundary for A. Thus $\operatorname{Ch}(A) \subset S(A)$. Since $S(A)$ is closed in X, it follows that the closure of $\operatorname{Ch}(A)$ is also contained in $S(A)$.

(c) If $\operatorname{Re} A$ separates the points of X/τ, then by part (b) of Corollary 4.1.11, $\operatorname{Ch}(A)$ is a Choquet set for A. By (b) above, it is contained in $S(A)$. Since $\operatorname{Ch}(A) \neq \phi$, we see that $S(A) \neq \phi$. $\square$

Example 4.2.3 Theorem 4.2.2 implies that for a (real) subspace A of $C(X,\tau)$, every closed boundary is a Choquet set; we give an example to show that a closed Choquet set may not be a boundary. Let $X = [0,1]$, $\tau(x) = 1 - x$ and $f(x) = i(1 - 2x)$. Then $f \in C(X,\tau)$. Let $A := \{s + tf : s,t \in \mathbb{R}\}$. Now, $\operatorname{Re} A = \{\text{real constant functions}\}$ and $\{|f|^2 : f \in A\} = \{s^2 + t^2u^2 : s,t \in \mathbb{R}\}$, where $u(x) = 1 - 2x$, $x \in X$. Hence every τ-invariant subset of $[0,1]$ is a Choquet set for A, while $\{0,1\}$ is the smallest boundary for A. (*Note*: u^2 assumes its maximum at 0 and 1 only.) Thus if $0 < x_0 < 1$, then $\{x_0, 1 - x_0\}$ is a closed Choquet set for A, but is not a boundary for A. Notice that in this example $\operatorname{Re} A$ does not separate the points of X/τ. Also, note that $\operatorname{Ch}(A) = \phi$, since every probability measure on X is an RPR measure for every $\operatorname{Re} e_x, x \in X$.

Next, let $A := \{s + te^f : s,t \in \mathbb{R}\}$. Then $\operatorname{Re} A = \{s + t \cos u : s,t \in \mathbb{R}\}$ and

$$\{|g|^2 : g \in A\} = \{s^2 + t^2 + 2st \cos u : s,t \in \mathbb{R}\}.$$

Since $\cos u = \cos(1 - 2x)$ assumes its maximum at $\frac{1}{2}$, and its minimum at 0 and 1, it follows that $\{0,\frac{1}{2},1\}$ is the smallest Choquet set for A and also the smallest boundary for A. Clearly, $\operatorname{Ch}(A) = \{0, \frac{1}{2}, 1\} = S(A)$. Note that in this example, $\operatorname{Re} A$ separates the points of X/τ.

Now we turn our attention to uniformly closed (real) subalgebras of

$C(X,\tau)$ containing 1. In addition to the characterizations of $Ch(A)$ given in Theorem 4.1.5, we have the following results.

Theorem 4.2.4 Let A be a uniformly closed (real) subalgebra of $C(X,\tau)$ containing 1. Let $x \in X$. Then the following statements are equivalent:

(i) $x \in Ch(A)$.

(ii) For every γ and δ with $0 < \gamma < \delta < 1$ and for every τ-invariant neighborhood U of $\{x,\tau(x)\}$, there exists $g \in A$ with $\|g\| \le 1$, $|g(x)| > \delta$, and $|g(y)| < \gamma$ for all y in $X \backslash U$.

(iii) There exist c and M with $0 < c < 1 \le M$ such that for every neighborhood V of $\{x,\tau(x)\}$, there exists g_1 in A with $\|g_1\| \le M$, $g_1(x) = 1$, and $|g_1(y)| < c$ for all y in $X \backslash V$.

(iv) $\{x,\tau(x)\}$ is a weak peak set for A.

(v) A is extremely regular at $\{x,\tau(x)\}$.

Proof.

(i) implies (ii): Let $\alpha = -\ln \delta$, $\beta = -\ln \gamma$, and $g = \exp(f)$, where f is as given in statement (iii) of Theorem 4.1.5.

(ii) implies (iii): Let $0 < \delta < 1$, and choose γ such that $0 < \gamma < \sqrt{1 + \delta^2} - 1$. Then $\gamma < \delta$. Let V be a neighborhood of $\{x,\tau(x)\}$, $U := V \cap \tau(V)$. Then U is a τ-invariant neighborhood of $\{x,\tau(x)\}$. By (ii), there is g in A such that $\|g\| \le 1$, $|g(x)| > \delta$ and $|g(y)| < \gamma$ for all y in $X \backslash U$. If $g(x) = a + ib$, define

$$g_1 := \frac{g(2a - g)}{a^2 + b^2}.$$

Then $g_1 \in A$. Since $|a| \le |g(x)| \le 1$ and $a^2 + b^2 = |g(x)|^2 \ge \delta^2$, we see that $\|g_1\| \le 3/\delta^2$, $g_1(x) = (a + ib)(a - ib)/(a^2 + b^2) = 1$ and $|g_1(y)| < \gamma(2 + \gamma)/\delta^2$ for all y in $X \backslash U$. Since $1 + \gamma < \sqrt{1 + \delta^2}$ and $X \backslash V \subset X \backslash U$, (iii) holds with $c = \gamma(2 + \gamma)/\delta^2 < 1$ and $M = 3/\delta^2 \ge 1$.

(iii), (iv), and (v) are equivalent by Theorem 2.2.8.

(v) implies (i) by Corollary 4.1.7. $\qquad\square$

The case $\gamma = \frac{1}{4}$ and $\delta = \frac{3}{4}$ of statement (ii) of Theorem 4.2.4 is known as Bishop's $\frac{1}{4}$–$\frac{3}{4}$ criterion.

Theorem 4.2.5 Let A be a uniformly closed (real) subalgebra of $C(X,\tau)$ containing 1.

(a) A subset of X is a boundary for A if and only if it is a Choquet set for A.

(b) If Re A separates the points of X/τ, then Ch(A) is a boundary for A and its closure equals $S(A)$.

Proof.

(a) Let S be a boundary for A. Since for every f in A, $\exp(f) \in A$, and $\operatorname{Re} f = \ln|\exp(f)|$, it follows that $\operatorname{Re} f$ assumes its maximum on S. Hence S is a Choquet set for A.

Conversely, let S be a Choquet set for A. If it were not a boundary for A, there is f in A with $\|f\| = 1$, $|f(x)| < 1$ for all x in S, but $|f(y)| = 1$ for some y in $X\backslash S$. Since $-\operatorname{Re} f(x) \leq |f(x)| < 1$ for all x in S and $\operatorname{Re}(-f)$ assumes its maximum on S, we see that $-\operatorname{Re} f(x) < 1$ for all x in X. By the continuity of $\operatorname{Re} f$ on the compact space X, there exists $\delta < 1$ such that $-\operatorname{Re} f \leq \delta$.

Consider $\phi \in \operatorname{Car}(A)$. By Lemma 1.2.2, $\|\phi\| = 1 = \phi(1)$. It is easy to see that $\operatorname{Re} \phi \in K(A)$. Since $\operatorname{Re}(f + \delta) \geq 0$, it follows by Lemma 3.2.4 that $\operatorname{Re} \phi(f + \delta) \geq 0$; that is, $-\delta \leq \operatorname{Re} \phi(f)$. [An alternative proof of this fact, which does not use Lemma 3.2.4, can be given as follows: Since ϕ is multiplicative and $\|\phi\| = 1$,

$$-\operatorname{Re} \phi(f) = \ln|\exp(\phi(-f))| = \ln|\phi(\exp(-f))|$$
$$\leq \ln\|\exp(-f)\| = \ln \sup_{x \in X} \exp(\operatorname{Re}(-f))$$
$$\leq \ln(\exp(\delta)) = \delta.]$$

As $\delta < 1$, it follows that $0 < 1 - \delta \leq \operatorname{Re} \phi(f + 1)$, and so $\phi(f + 1) \neq 0$. This is true for every $\phi \in \operatorname{Car}(A)$. By (v) of Theorem 1.2.9, we see that $f + 1$ is invertible in A. Let $g := (f - 1)(f + 1)^{-1}$. Then $g \in A$ and for x in X,

$$\operatorname{Re} g(x) = \frac{|f(x)|^2 - 1}{|f(x) + 1|^2}.$$

For x in S, $|f(x)| < 1$ and hence $\operatorname{Re} g(x) < 0$, while $|f(y)| = 1$ and hence $\operatorname{Re} g(y) = 0$. Thus $\operatorname{Re} g$ does not assume its maximum on S, a contradiction.

(b) Let $\operatorname{Re} A$ separate the points of X/τ. Then by part (b) of Corollary 4.1.11, Ch(A) is a Choquet set, and hence by (a) above, it is a boundary for A. Since by definition, $S(A)$ is the intersection of all closed boundaries for A, it is contained in the closure of Ch(A) (which is a closed boundary). We have already seen in part (b) of Theorem 4.2.2 that $S(A)$ contains the closure of Ch(A). Hence all is proven. □

Theorem 4.2.5 allows us to make the following nomenclature.

Definition 4.2.6 Let A be a uniformly closed (real) subalgebra of $C(X,\tau)$ containing 1 such that Re A separates the points of X/τ. The set Ch(A) is called the *Choquet boundary for A*, and its closure $S(A)$ is called the *Shilov boundary* for A.

Theorem 4.2.5 and Definition 4.2.6 apply when A is a real function algebra on (X,τ), since it is a uniformly closed (real) subalgebra of $C(X,\tau)$, it contains constants and Re A separates the points of X/τ by Lemma 1.3.9(iii).

Example 4.2.7 If A is a real function algebra on (X,τ) and Re A is uniformly dense in $\{u \in C_{\mathbb{R}}(X): u \circ \tau = u\}$, then by Theorem 4.1.3, Ch(A) = X and hence $S(A) = X$. In particular, Ch($C(X,\tau)$) = $S(C(X,\tau))$ = X.

Next, let A be the real disk algebra of Example 1.3.10. As we have seen in Example 2.2.12, for each $s \in \mathbb{C}$ with $|s| = 1$, the set $\{s,\bar{s}\}$ is a (weak) peak set for A. On the other hand, if $s \in \mathbb{C}$ and $|s| < 1$, then A is not extremely regular at $\{s,\bar{s}\}$. For if $f \in A$ and $f(s) = 1 = \|f\|$, then by the maximum modulus theorem for analytic functions, f is the constant function 1. Hence by Theorem 4.2.4, Ch(A) = $\{s \in \mathbb{C}: |s| = 1\}$. Since in this case Ch($A$) is closed, it equals $S(A)$ by part (b) of Theorem 4.2.5.

4.3 CHOQUET SETS AND BOUNDARIES FOR COMPLEX SUBSPACES OF $C(X)$

Throughout this section, B will be a complex subspace of $C(X)$ containing 1.

Definition 4.3.1 A subset S of X is called a *Choquet set* (respectively, *boundary*) for B if Re h (respectively, $|h|$) assumes its maximum on S for every h in B.

As in Theorem 4.2.2(a), if S is a boundary for B, then the closure of S is a Choquet set for B. In fact, the proof of Theorem 4.2.2(a) is valid if B is merely a real subspace of $C(X)$ containing 1. Also, if B is a real or complex subalgebra of $C(X)$ containing 1, then every boundary for B is a Choquet set for B, as the proof of Theorem 4.2.5(a) shows. On the other hand, we have seen in Example 4.2.3 a real subspace of $C(X,\tau)$ containing 1 for which a closed Choquet set is not a boundary. However, if B is a complex subspace of $C(X)$, then every Choquet set S for B is a boundary for B. This can be seen as follows: Let $0 \neq h \in B$. Then there is $x \in X$ such that $0 \neq |h(x)| = \|h\|_\infty$. Now, $h_1 = h/h(x)$ is in B and $\|h_1\|_\infty = 1$. Since

Re $h_1 \le |h_1| \le 1$ on X, Re $h_1(x) = 1$ and S is a Choquet set for B, there is $x_1 \in S$ such that Re $h_1(x_1) = 1$. But then $1 = $ Re $h_1(x_1) \le |h_1(x_1)| \le 1$, so that $|h_1(x_1)| = 1 = \|h_1\|_\infty$, and in turn $|h(x_1)| = \|h\|_\infty$. This shows that S is a boundary for B. Let

$$S(B) = \cap\{S : S \text{ is a closed boundary for } B\}.$$

Then, by what we have just shown, $S(B)$ is also the intersection of all closed Choquet sets for B.

Recalling Definition 3.2.2 of a representing measure for $\phi \in B'$ and Theorem 3.2.3(a), we let

$$\text{Ch}(B) = \{x \in X : e_x \text{ has a unique representing measure on } X\}.$$

For $x \in X$, the measure δ_x is a representing measure for e_x. If $x \in \text{Ch}(B)$, then δ_x is the only representing measure for e_x.

By suitably modifying the proofs of results in Sections 4.1 and 4.2, analogous results can be obtained for Choquet sets and boundaries for B and for $\text{Ch}(B)$. We refrain from working out all the details, but only state some major results. Some of these can be found in Browder (1969, Sections 2.2 and 2.3).

Recall that

$$K(B) = \{\phi \in B' : \|\phi\| = 1 = \phi(1)\}.$$

For $\phi \in K(B)$ and $u \in C_{\mathbb{R}}(X)$, let

$$\alpha_u(\phi) := \sup\{\text{Re } \phi(h) : h \in B, \quad \text{Re } h \le u\}$$

and

$$\beta_u(\phi) := \sup\{\text{Re } \phi(h) : h \in B, \quad \text{Re } h \ge u\}.$$

Then by Lemma 3.2.4, $\alpha_u(\phi) \le \beta_u(\phi)$.

Theorem 4.3.2 Let $x \in X$. The following statements are equivalent:
(i) $x \in \text{Ch}(B)$.
(ii) $\alpha_u(e_x) = u(x) = \beta_u(e_x)$ for all $u \in C_{\mathbb{R}}(X)$.
(iii) For every α and β with $0 < \alpha < \beta$ and for every neighborhood U of x, there exists h in B with Re $h \le 0$, Re $h(x) > -\alpha$ and Re $h(y) < -\beta$ for all y in $X\backslash U$.
(iv) There exist α and β with $0 < \alpha < \beta$ such that for every neighborhood U of x, there exists h in B with Re $h \le 0$, Re $h(x) > -\alpha$, and Re $h(y) < -\beta$ for all y in $X\backslash U$.
(v) If μ is a representing measure for e_x, then $\mu(\{x\}) > 0$.
As a consequence,

$$\text{Ch}(B) \subset S(B).$$

Proof. The proof of the equivalence of statements (i) to (v) is similar to that of Theorem 4.1.5. Again, as in Corollary 4.1.6, Ch(B) is contained in every closed Choquet set for B, and hence in $S(B) = \cap\{S : S$ is a closed Choquet set for $B\}$, as pointed out in the comment after Definition 4.3.1. □

Theorem 4.3.3 If $x \in$ Ch(B), then e_x is an extreme point of $K(B)$. Conversely, if ϕ is an extreme point of $K(B)$, then $\phi = e_x$ for some $x \in X$, and if B separates x from other points of X, then, in fact, $x \in$ Ch(B).

Proof. Similar to that of Theorem 4.1.10. □

Corollary 4.3.4 Assume that B separates the points of X. Then the following statements hold.
(a) Ch(B) $= \{x \in X : e_x$ is an extreme point of $K(B)\}$.
(b) Ch(B) is a Choquet set for B, and in particular, a boundary for B.
(c) The closure of Ch(B) equals $S(B)$, the smallest closed boundary for A.

Proof. Statement (a) follows from Theorem 4.3.3. The proof of (b) is similar to that of Corollary 4.1.11(b), if we let

$$L := \{\phi \in K(B) : \mathrm{Re}\, \phi(f) = \mathrm{Re}\, f(x)\},$$

where $x \in X$ with $\mathrm{Re}\, f(x) = \max\{\mathrm{Re}\, f(y) : y \in X\}$ and apply Lemma 3.2.4 to $\phi \in K(B)$. Finally, (c) follows since by Theorem 4.3.2, the closure of Ch(B) is contained in $S(B)$ and is itself a closed boundary for B. □

Definition 4.3.5 Let B be a complex subspace of $C(X)$ containing 1 and separating the points of X. The set Ch(B) is called the *Choquet boundary for B* and its closure $S(B)$ is called the *Shilov boundary for B*.

Theorem 4.3.6 Let B be a uniformly closed complex subalgebra of $C(X)$ containing 1. Let $x \in X$. Then the following statements are equivalent:
(i) $x \in$ Ch(B).
(ii) For every γ and δ with $0 < \gamma < \delta < 1$ and for every neighborhood U of x, there exists $k \in B$ with $\|k\| \leq 1$, $k(x) > \delta$, and $|k(y)| < \gamma$ for all y in $X\backslash U$.
(iii) There exist c and M with $0 < c < 1 \leq M$ such that for every neighborhood V of x, there exists k_1 in B with $\|k_1\| \leq M$, $k_1(x) = 1$, and $|k_1(y)| < c$ for all y in $X\backslash V$.
(iv) $\{x\}$ is a weak peak set for B.
(v) B is extremely regular at $\{x\}$.

Proof. (i) implies (ii) by letting $\alpha = -\ln \delta$, $\beta = -\ln \gamma$, and $k = \exp(h - i \operatorname{Im} h(x))$, where h is as given in statement (iii) of Theorem 4.3.2. (ii) implies (iii) by letting $c = \gamma/\delta$, $M = 1/\delta$, and $k_1 = k/k(x)$, where $0 < \gamma < \delta < 1$ and k is as given in (ii). (iii), (iv), and (v) are equivalent by Theorem 2.2.8.

(v) implies (i): For every neighborhood U of x, there is h_1 in B such that $\|h_1\|_\infty \leq 1$, $h_1(x) = 1$, and $|h_1(y)| < \frac{1}{4}$ for every y in $X \backslash U$. Let $h := h_1 - 1$, $\alpha = \frac{1}{4}$, and $\beta = \frac{3}{4}$. Then $h \in B$, $0 < \alpha < \beta$, $\operatorname{Re} h \leq 0$, $\operatorname{Re} h(x) > -\alpha$, and $\operatorname{Re} h(y) < -\beta$ for all y in $X \backslash U$. Hence $x \in \operatorname{Ch}(B)$ by statement (iv) of Theorem 4.3.2. □

We note that since multiplication by complex scalars is permissible in B, the proof of Theorem 4.3.6 is a lot simpler than the proof of the corresponding Theorem 4.2.4 for real subalgebras of $C(X,\tau)$.

Theorems 4.3.2, 4.3.3, Corollary 4.3.4, and Theorem 4.3.6 are valid for a complex function algebra B on X, since B is a complex subspace of $C(X)$, contains 1, separates the points of X, and is, in fact, a uniformly closed complex subalgebra of $C(X)$.

In case a complex subspace B of $C(X)$ is a "complexification" of a (real) subspace A of $C(X,\tau)$ containing 1, we wish to relate $\operatorname{Ch}(B)$ to $\operatorname{Ch}(A)$ and $S(B)$ to $S(A)$.

Theorem 4.3.7 Let A be a (real) subspace of $C(X,\tau)$ containing 1 and let

$$B = \{f + ig : f, g \text{ in } A\} \subset C(X).$$

(a) If A separates x and $\tau(x)$ whenever they are distinct, then $\operatorname{Ch}(A) \subset \operatorname{Ch}(B)$.
(b) If A is a uniformly closed subalgebra, then $\operatorname{Ch}(B) \subset \operatorname{Ch}(A)$.
(c) If A is a real function algebra on (X,τ), then $\operatorname{Ch}(A) = \operatorname{Ch}(B)$ and $S(A) = S(B)$.

Proof.
(a) Let $x \in \operatorname{Ch}(A)$. Consider a representing measure μ for $e_x : B \to \mathbb{C}$. To show $x \in \operatorname{Ch}(B)$, we must conclude that $\mu = \delta_x$.

Since μ is a probability measure, so is $\nu := (\mu + \mu \circ \tau)/2$. Also, for f in A, we have

$$\operatorname{Re} f(x) = \frac{f(x) + \bar{f}(x)}{2} = \frac{\int_X f \, d\mu + \int_X \bar{f} \, d\mu}{2}$$

$$= \frac{\int_X f\, d\mu + \int_X f \circ \tau\, d\mu}{2} = \frac{\int_X f\, d\mu + \int_X f\, d\mu \circ \tau}{2}$$

$$= \int_X f\, d\nu = \int_X \operatorname{Re} f\, d\nu,$$

since $\nu \circ \tau = \nu$. This shows that ν is a RPR measure for $\operatorname{Re} e_x$. As $x \in \operatorname{Ch}(A)$, we see that $\nu = m_x$, that is,

$$\frac{\mu + \mu \circ \tau}{2} = \frac{\delta_x + \delta_{\tau(x)}}{2}.$$

This implies that the support of the probability measure μ is contained in $\{x, \tau(x)\}$, so that $\mu = t\delta_x + (1 - t)\delta_{\tau(x)}$ for some $0 \le t \le 1$. If $\tau(x) = x$, then clearly $\mu = \delta_x$, as desired. Now, let $\tau(x) \ne x$. For f in A,

$$f(x) = \int_X f\, d\mu = tf(x) + (1 - t)f(\tau(x)),$$

that is, $(1 - t)f(x) = (1 - t)f(\tau(x))$. If A separates x and $\tau(x)$, that is, $f(x) \ne f(\tau(x))$ for some f in A, we see that $t = 1$, yielding $\mu = \delta_x$.

(b) Let $x \in \operatorname{Ch}(B)$. If A is a uniformly closed (real) subalgebra of $C(X, \tau)$ containing 1, then by Theorem 1.3.20, B is a uniformly closed complex subalgebra of $C(X)$ containing 1. Hence by Theorem 4.3.6, $\{x\}$ is a weak peak set for B. But then Theorem 2.2.11 shows that $\{x, \tau(x)\}$ is a weak peak set for A. Now, Theorem 4.2.4 implies that $x \in \operatorname{Ch}(A)$.

(c) If A is a real function algebra on (X, τ), then A separates the points of X, so that $\operatorname{Ch}(A) \subset \operatorname{Ch}(B)$ by (a), and A is a uniformly closed subalgebra of $C(X, \tau)$, so that $\operatorname{Ch}(B) \subset \operatorname{Ch}(A)$ by (b). Thus $\operatorname{Ch}(A) = \operatorname{Ch}(B)$. Since Lemma 1.3.9(ii) shows that $\operatorname{Re} A$ separates the points of X/τ, the closure of $\operatorname{Ch}(A)$ equals $S(A)$ according to Theorem 4.2.5(b). Also, since B separates the points of X, the closure of $\operatorname{Ch}(B)$ equals $S(B)$ according to Corollary 4.3.4. Hence $S(A) = S(B)$.   □

Remark 4.3.8 The assumption in part (a) of Theorem 4.3.7 that A separates x and $\tau(x)$ whenever they are distinct cannot be dropped. Let τ be an involution on X and $A := \{u \in C_{\mathbb{R}}(X) : u \circ \tau = u\}$. Then A is a (real) subspace (in fact, a uniformly closed subalgebra) of $C(X, \tau)$ containing 1. Since $\operatorname{Re} A = \{u \in C_{\mathbb{R}}(X) : u \circ \tau = u\}$, it follows by Theorem 4.1.3 that $\operatorname{Ch}(A) = X$. Next, $B = \{u + iv : u, v \in C_{\mathbb{R}}(X), u \circ \tau = u, v \circ \tau = v\}$. Let $x \in X$ with $\tau(x) = x$. Let μ be a representing measure for $e_x : B \to \mathbb{C}$. Then for every $u \in C_{\mathbb{R}}(X)$ with $u \circ \tau = u$,

$$\int_X u \, d\mu = u(x) = \int_X u \, d(\mu \circ \tau).$$

If we let $v = (\mu + \mu \circ \tau)/2$, then we have

$$\int_X u \, dv = \int_X u \, d\delta_x$$

for every $u \in C_\mathbb{R}(X)$ with $u \circ \tau = u$, where $v \circ \tau = v$ and $\delta_x \circ \tau = \delta_x$. Hence by Lemma 4.1.2, $v = \delta_x$, and in turn, $\mu = \delta_x$, showing that $x \in$ Ch(B). On the other hand, if $x \in X$ with $\tau(x) \neq x$, then δ_x and $\delta_{\tau(x)}$ are distinct representing measures for $e_x : B \to \mathbb{C}$, so that $x \notin$ Ch(B). Thus

$$\text{Ch}(B) = \{x \in X : \tau(x) = x\}.$$

If τ is not the identity map on X, it follows that $X = $ Ch(A) is not contained in Ch(B).

Similarly, the assumption in part (b) of Theorem 4.3.7 that A is a subalgebra of $C(X,\tau)$ cannot be dropped. Let $X = [0,1]$, $\tau(x) := 1 - x$ for $x \in X$. Consider $f(x) := i(1 - 2x)$, $x \in X$. Then $f \in C(X,\tau)$. If $A := \{s + tf : s,t \in \mathbb{R}\}$, then A is a (real) subspace of $C(X,\tau)$ containing 1 (and separating x and $\tau(x)$ whenever they are distinct). We have seen in Example 4.2.3 that Ch(A) = ϕ. Next,

$$B = \{z + wf : z,w \in \mathbb{C}\} = \{z + wg : z,w \in \mathbb{C}\},$$

where $g(x) = x, x \in X$. B is thus a complex subspace of $C(X)$ containing 1 and separating the points of X. By Corollary 4.3.4(b), Ch(B) is a Choquet set for B. Since Re g and Re$(1 - g)$ attain their maxima only at 1 and 0, respectively, it follows that $\{0,1\}$ is contained in every Choquet set; in particular, $\{0,1\} \subset$ Ch(B). On the other hand, since Re $B = \{s + tg : s,t \in \mathbb{R}\}$, it can be seen that $\{0,1\}$ is a Choquet set for B. Hence Ch(B) $\subset \{0,1\}$ by Theorem 4.3.2. Thus Ch(B) $= \{0,1\}$.

Example 4.3.9 If for a real function algebra A on (X,τ), we have Ch(A) $= X$; then for its complexification B we have Ch(B) = Ch(A) = $X = S(A) = S(B)$. In particular, we have Ch($C(X)$) $= X = S(C(X))$, since Ch($C(X,\tau)$) $= X$, as seen in Example 4.2.7.

We have also seen in Example 4.2.7 that if A is the real disk algebra, then

$$\text{Ch}(A) = \{s \in \mathbb{C} : |s| = 1\} = S(A).$$

Since the complex disk algebra B is a complexification of A (Example 1.3.10), it follows from Theorem 4.3.7 that

$$\mathrm{Ch}(B) = \{s \in \mathbb{C} : |s| = 1\} = S(B).$$

Of course, this can also be seen independently, by noting that every $s \in \mathbb{C}$ with $|s| = 1$ is a (weak) peak set for B since the function $f(z) = (1 + \bar{s}z)/2$ in B peaks at s, and by using the maximum modulus theorem for analytic functions to conclude that B cannot be extremely regular at $\{s\}$ if $|s| < 1$.

Next, we turn our attention to the real algebra A_q introduced in Example 3.3.6. Let C be a complex function algebra on $Z = \mathrm{Car}(C)$. Let $Z_q = \{z_1, \ldots, z_q\}$ be a specified finite subset of Z and for $1 \le k \le q$, and let D_k be a continuous point derivation of C at z_k (recall Definition 3.3.5). Let

$$A_q := \{f \in C : f(z_k) \text{ and } D_k(f) \text{ are real for } 1 \le k < q\}.$$

Let X denote the union of two copies of Z identified at Z_q. Let $\tau: X \to X$ be the map that sends a point in one copy of Z to the corresponding point in the other copy. In particular, $\tau(x) = x$ if and only if $x = z_k$ for some k, $1 \le k \le q$. Then A_q can be regarded as a real function algebra on (X, τ) with $X = \mathrm{Car}(A_q)$. We wish to compute $\mathrm{Ch}(A_q)$ if we are given $\mathrm{Ch}(C)$. For this purpose we first prove a preliminary result.

Lemma 4.3.10 Let B be a uniformly closed complex subalgebra of $C(X)$ containing 1, and $x \in \mathrm{Ch}(B)$. Then every continuous point derivation of B at e_x is zero.

Proof. Let D be a continuous point derivation of B at e_x. Since $D(1) = 0$, it follows that for every f in B, $D(f) = D(f - f(x)) + D(f(x)) = D(f - f(x)) + f(x)D(1) = D(f - f(x))$. Thus it suffices to prove that $D(f) = 0$, whenever $f \in B$ and $f(x) = 0$. We may assume, in addition, that $\|f\| = 1$. For $n = 1, 2, \ldots$, let $U_n := \{y \in X : |f(y)| < 1/n\}$. Since $x \in \mathrm{Ch}(B)$, B is extremely regular at $\{x\}$ by Theorem 4.3.6. Hence there is $g_n \in B$ such that $g_n(x) = 1 = \|g_n\|$ and $|g_n(y)| < 1/n$ for all y in $X \setminus U_n$. Then it follows that $\|fg_n\|_\infty \le 1/n$. Now, for each n,

$$D(fg_n) = f(x)D(g_n) + D(f)g_n(x) = D(f),$$

since $f(x) = 0$ and $g_n(x) = 1$. As D is continuous, we see that $|D(f)| = |D(fg_n)| \le \|D\|/n \to 0$; that is, $D(f) = 0$. ☐

Theorem 4.3.11 $\mathrm{Ch}(A_q)$ is the union of two copies of $\mathrm{Ch}(C)$ identified at $Z_q \cap \mathrm{Ch}(C)$, and $S(A_q)$ is the union of two copies of $S(C)$ identified at $Z_q \cap S(C)$.

Proof. Let $x \in \mathrm{Ch}(A_q) \subset X$. Then by Theorem 4.2.4, there are constants c and M with $0 < c < 1 \le M$ such that for every neighborhood V of $\{x, \tau(x)\}$, there exists g_1 in A_q with $\|g_1\| \le M$, $g_1(x) = 1$, and $|g_1(y)| < c$

for all y in $X \backslash V$. Since X is the union of Z and $\tau(Z)$ identified at Z_q, either $x \in Z$ or $\tau(x) \in Z$. Assume first that $x \in Z$, and let W be a neighborhood of x in Z. Then $V := W \cup \tau(W)$ is a neighborhood of $\{x, \tau(x)\}$ in X. Hence there is $g_1 \in A_q$ such that $\|g_1\| \le M$, $g_1(x) = 1$, and $|g_1(y)| < c$ for all y in $X \backslash V$. Since $A_q \subset C$ and $Z \backslash W \subset X \backslash V$, it follows again from Theorem 4.2.4 that $x \in \text{Ch}(C)$. Similarly, if $\tau(x) \in Z$, then we can show that $\tau(x) \in \text{Ch}(C)$; that is, $x \in \tau(\text{Ch}(C))$. This shows that $\text{Ch}(A_q)$ is contained in the union E of two copies of $\text{Ch}(C)$ identified at $Z_q \cap \text{Ch}(C)$.

Conversely, let $x \in E$. Assume first that $x \in \text{Ch}(C) \subset Z$. To show $x \in \text{Ch}(A_q)$, we first construct a function f_0 in C such that

$$f_0(x) = 1, \quad f_0(z_k) = 0 \quad \text{if } z_k \ne x, \quad k = 1, \ldots, q,$$

and

$$D_k(f_0) = 0, \quad k = 1, \ldots, q.$$

Case (i): Suppose that $x \ne z_k$ for $k = 1, \ldots, q$. Let D_0 denote the zero point derivation at x. By Lemma 3.3.7(i), there is $f_k \in C$ such that $f_k(x) = 1$, $f_k(z_k) = 0$, and $D_0(f_k) = 0 = D_k(f_k)$, for $1 \le k \le q$. Let $f_0 = f_1 \cdots f_q$. If $q = 1$, then obviously $f_0 = f_1$, works. If $q \ge 2$, then again it is clear that for every k, $1 \le k \le q$, $f_0(z_k) = 0$, and

$$D_k(f_0) = \left(\prod_{j=1, j \ne k}^{q} f_j \right)(z_k) D_k(f_k) + f_k(z_k) D_k \left(\prod_{j=1, j \ne k}^{q} f_j \right)$$

$$= 0,$$

since $D_k(f_k) = 0 = f_k(z_k)$.

Case (ii): Suppose that $x = z_k$ for some k, say $x = z_1$. If $q = 1$, let $f_0 = 1$. If $q \ge 2$, again by Lemma 3.3.7(i) there is $f_k \in C$ such that $f_k(z_1) = 1$, $f_k(z_k) = 0$, and $D_1(f_k) = 0 = D_k(f_k)$, for $2 \le k \le q$. Let $f_0 = f_2 \cdots f_q$. Then $f_0(x) = f_0(z_1) = 1$, and for every k, $2 \le k \le q$, $f_0(z_k) = 0 = D_k(f_0)$ as in case (i). Also, $D_1(f_0) = 0$ as in the proof of Lemma 3.3.7(ii). [Alternatively, since $z_1 = x \in \text{Ch}(C)$, $D_1 = 0$ by Lemma 4.3.10, so that $D_1(f_0) = 0$.]

Now since $x \in \text{Ch}(C)$, Theorem 4.3.6 shows that C is extremely regular at $\{x\}$. Hence for every neighborhood W of x in Z, and every $\varepsilon > 0$, there is f in C such that $f(x) = 1 = \|f\|$ and $|f(y)| < \varepsilon$ for all y in $Z \backslash W$. Consider now a τ-invariant neighborhood V of $\{x, \tau(x)\}$ in X. Let $W = V \cap Z$, $\varepsilon = 1/(2\|f_0\|)$, and $g = f_0 f$, where f_0 and f are as described above. Then clearly $g(x) = 1$, $\|g\| \le \|f_0\|$, and $|g(y)| < \frac{1}{2}$ for all y in $Z \backslash W$, and hence in $X \backslash V$. Also, in case (i), $g(z_k) = 0$ and $D_k(g) = f_0(z_k) D_k(f) + f(z_k) D_k(f_0) = 0$ for $k = 1, \ldots, q$ and in case (ii), $g(z_1) = 1$, $D_1(g) = 0$ by Lemma 4.3.10 and $g(z_k) = 0 = D_k(g)$ for $k = 2, \ldots, q$. Hence $g \in A_q$.

Thus statement (iii) of Theorem 4.2.4 is satisfied with $c = \frac{1}{2}$ and $M = \|f_0\|$. It follows that $x \in \text{Ch}(A_q)$. A similar proof shows that if $\tau(x) \in \text{Ch}(C)$, then $\tau(x) \in \text{Ch}(A_q)$ and since $\text{Ch}(A_q)$ is τ-invariant, we again have $x \in \text{Ch}(A_q)$. Thus the union E of two copies of $\text{Ch}(C)$ identified at $Z_q \cap \text{Ch}(C)$ is contained in $\text{Ch}(A_q)$.

Since $S(A_q)$ [respectively $S(C)$] is the closure of $\text{Ch}(A_q)$ [respectively $\text{Ch}(C)$] in X (respectively Z), we see that $S(A_q)$ is the union of two copies of $S(C)$, identified at $Z_q \cap S(C)$. $\qquad\qquad\square$

Remark 4.3.12 Before we conclude this chapter we mention that several authors have previously studied the concept of a minimal closed boundary for a real commutative Banach algebra A with identity 1. Ingelstam (1968) was perhaps the first to define the Shilov boundary for A as the smallest closed τ-invariant subset of $\text{Car}(A)$ on which $|\hat{f}|$ assumes its maximum for each f in A, where $\hat{f}$ is the Gelfand transform of f. Limaye and Simha (1975) define a boundary for A as a subset of the maximal ideal space $M(A)$ of A on which $|\hat{f}|$ assumes its maximum for each f in A, where $|\hat{f}|$ is as in Remark 1.2.11. They show that if B is the complexification of A, $cx^*: M(B) \to M(A)$ is the restriction map and $\text{SL}(B) := \{e_x^{-1}(0): x \in S(B)\}$, then $\text{SL}(A) := cx^*(\text{SL}(B))$ is the smallest closed boundary for A. This result can also be deduced from Theorem 4.3.7. Srivastav (1991) proves it using spectral states, even when A does not have a unit but $M(A) \neq \phi$. Limaye (1976) defines a Choquet set for A as a subset of $M(A)$ on which $\text{Re}\,\hat{f}$ (see Remark 1.2.11) assumes its maximum for each f in A and shows that (i) every boundary for A is a Choquet set for A, and (ii) every closed Choquet set for A is a boundary for A. This result can be improved in view of Theorem 4.2.5(a). This paper also contains the following results. Let A_0 be a closed (real) subalgebra of A containing 1, and let $r: M(A) \to M(A_0)$ be the restriction map. Then $\text{SL}(A_0) \subset r(\text{SL}(A))$. If r is 1:1 on $\text{SL}(A)$, then $\text{SL}(A) \cap r^{-1}(\text{SL}(A_0))$ is the smallest closed subset of $\text{SL}(A)$ on which $|\hat{f}|$ attains its maximum for every f in A_0. Conditions are also given under which $\text{SL}(A_0) = r(\text{SL}(A))$.

Limaye, Mehta, and Vasavada (1982) have shown that if A_1 and A_2 are real Banach algebras with units, then there exists an at most $2:1$ continuous open map from the maximal ideal space of the tensor product of A_1 and A_2 (with the greatest cross norm) to $M(A_1) \times M(A_2)$, which maps $\text{SL}(A_1 \otimes A_2)$ onto $\text{SL}(A_1) \times \text{SL}(A_2)$.

Concerning the real algebra A_q discussed in Theorem 4.3.11, Limaye and Simha (1975) prove that $cx^*(\text{SL}(C))$ is the smallest closed boundary for A_q under the hypothesis that $\text{Re}\,C$ has finite codimension in $C_{\mathbb{R}}(Z)$. This hypothesis is not necessary in view of Theorem 4.3.11 which appears in Kulkarni and Arundhathi (1988). This article also contains Theorems 4.1.5, 4.1.10, a version of 4.2.5, the last part of 4.3.7 and 4.3.11 for real function algebras. The result Theorem 4.2.4(v) appears in Kulkarni and Arundhathi (1991a). The proof of Theorem 4.3.7(b) given by us is from Hwang (1990). Most of the remaining results are proved in Kulkarni and Limaye (1991a).

5

Isometries of Real Function Algebras

If two real function algebras are linearly isometric (as Banach spaces), can we say that those algebras are isomorphic (as algebras)? This is the central question of the present chapter. We shall answer this question in the affirmative. The origin of such questions can be traced back to the classical Banach–Stone theorem, which states that if $C(X)$ and $C(Y)$ are linearly isometric, where X and Y are compact Hausdorff spaces, then X and Y are homeomorphic. This implies that $C(X)$ and $C(Y)$ are isomorphic. Several proofs of this theorem are available in the literature. We refer the reader to the surveys by Behrends (1979) and Jarosz (1985) for these proofs as well as for many generalizations of the Banach–Stone theorem. In this chapter we shall be concerned with one such generalization due to Nagasawa (1959). He proved that if two complex function algebras are linearly isometric, then those algebras are isomorphic. Our aim is to extend this result to the case of real function algebras.

In Section 5.1 we prove that a linear isometry between two real function algebras, which preserves the units, is an algebra isomorphism. It is proved in Section 5.2 that whenever two real function algebras are linearly isometric, it is possible to find a linear isometry between the two algebras which preserves the units. These two facts imply the extension of Nagasawa's theorem to real function algebras.

5.1 UNIT-PRESERVING ISOMETRIES

Recall the definition of an extremely regular subspace (Definition 2.2.6) and the consequences of extreme regularity proved in Lemma 2.2.7.

Lemma 5.1.1 Let X_1 and X_2 be compact Hausdorff spaces. For $j = 1,2$, let A_j be a real subspace of $C(X_j)$ which is extremely regular at a point x_j in X_j. Suppose that $T: A_1 \rightarrow A_2$ is a linear isometry of A_1 onto A_2 such that for all f in A_1

$$\mathrm{Re}((Tf)(x_2)) = \mathrm{Re}(f(x_1)).$$

Then $(Tf)(x_2) = f(x_1)$ for all f in A_1, or $(Tf)(x_2) = \overline{f(x_1)}$ for all f in A_1.

The same conclusion holds if for $j = 1,2$, τ_j is a topological involution on X_j, and A_j is a real subspace of $C(X_j, \tau_j)$ which is extremely regular at $\{x_j, \tau_j(x_j)\}$.

Proof.   First consider f in A_1 with $\mathrm{Re}\, f(x_1) \geq 0$ and $\|f\| = 1$. If $g \in A_1$ and $\mathrm{Re}\, g(x_1) = 1 = \|g\|$, then

$$\begin{aligned}
\|f + g\|^2 \geq |(f + g)(x_1)|^2 &= [\mathrm{Re}(f + g)(x_1)]^2 + [\mathrm{Im}(f + g)(x_1)]^2 \\
&= [\mathrm{Re}\, f(x_1) + 1]^2 + [\mathrm{Im}\, f(x_1)]^2 \\
&= |f(x_1)|^2 + 2\,\mathrm{Re}\, f(x_1) + 1 =: \alpha_f, \text{ say.}
\end{aligned}$$

Hence $\alpha_f \leq \inf\{\|f + g\|^2 : g \text{ in } A_1, \mathrm{Re}\, g(x_1) = 1 = \|g\|\}$. We show that, in fact, equality holds here. Let $\varepsilon > 0$. Since f is continuous at x_1, there is a neighborhood U of x_1 such that $|f(x) - f(x_1)| < \varepsilon$ for all x in U. As A_1 is extremely regular at x_1, there is g in A_1 such that $g(x_1) = 1 = \|g\|$, $|g(x)| < \varepsilon$ for all x in $X \backslash U$ and $|\mathrm{Im}\, g(x)| < \varepsilon$ for all x in X by Lemma 2.2.7.

Let $x \in U$. Then

$$\begin{aligned}
|f(x)|^2 \leq [|f(x_1)| + \varepsilon]^2 &= |f(x_1)|^2 + \varepsilon^2 + 2\varepsilon|f(x_1)| \\
&\leq |f(x_1)|^2 + \varepsilon^2 + 2\varepsilon,
\end{aligned}$$

since $|f(x_1)| \leq \|f\| = 1$. Also,

$$|\mathrm{Re}\, f(x) - \mathrm{Re}\, f(x_1)| \leq |f(x) - f(x_1)| < \varepsilon,$$

so that

$$|\mathrm{Re}\, g(x)[\mathrm{Re}\, f(x) - \mathrm{Re}\, f(x_1)]| < \varepsilon,$$

since $|\mathrm{Re}\, g(x)| \leq \|g\| = 1$, and hence

$$\mathrm{Re}\, g(x)\, \mathrm{Re}\, f(x) < \mathrm{Re}\, g(x)\, \mathrm{Re}\, f(x_1) + \varepsilon < \mathrm{Re}\, f(x_1) + \varepsilon$$

since $\mathrm{Re}\, f(x_1) \geq 0$. Now

$$|f(x) + g(x)|^2 = |f(x)|^2 + |g(x)|^2 + 2(\text{Re } f(x) \text{ Re } g(x)$$
$$+ \text{Im } f(x) \text{ Im } g(x))$$
$$\leq |f(x_1)|^2 + \varepsilon^2 + 2\varepsilon + 1 + 2 \text{ Re } f(x_1) + 2\varepsilon + 2\varepsilon$$
$$= \alpha_f + \varepsilon^2 + 6\varepsilon.$$

Let $x \in X \backslash U$. Then

$$|f(x) + g(x)|^2 = |f(x)|^2 + |g(x)|^2 + 2 \text{ Re } f(x)\overline{g(x)}$$
$$\leq 1 + \varepsilon^2 + 2\varepsilon \leq \alpha_f + \varepsilon^2 + 2\varepsilon.$$

If τ_1 is a topological involution on X_1 and A_1 is extremely regular at $\{x_1, \tau_1(x_1)\}$, we consider $V = U \cup \tau_1(U)$. Then by Lemma 2.2.7, there is g in A_1 such that $g(x_1) = g(\tau_1(x_1)) = 1 = \|g\|$, $|g(x)| < \varepsilon$ for all x in $X \backslash V$ and $|\text{Im } g(x)| < \varepsilon$ for all x in X.

Let $x \in V$. Then $x \in U$ or $x \in \tau_1(U)$. Let

$$x' = \begin{cases} x & \text{if } x \in U \\ = \tau(x) & \text{if } x \notin U. \end{cases}$$

As $f \in C(X_1, \tau_1)$, we have $|f(x)| = |f(x')|$ and since $x' \in U$, we can deduce, as above, that

$$|f(x)| = |f(x')| \leq \alpha_f + \varepsilon^2 + 6\varepsilon.$$

Let $x \in X \backslash V$. Then again the same argument as above shows that

$$|f(x) + g(x)|^2 \leq \alpha_f + \varepsilon^2 + 2\varepsilon.$$

Thus we see that in all the cases,

$$\|f + g\|^2 \leq \alpha_f + \varepsilon^2 + 6\varepsilon.$$

Since this is true for every $\varepsilon > 0$, we obtain

$$\alpha_f = \inf\{\|f + g\|^2 : g \in A_1, \text{ Re } g(x_1) = 1 = \|g\|\}.$$

Next, $\text{Re}(Tf)(x_2) = \text{Re } f(x_1) \geq 0$ and $\|Tf\| = \|f\| = 1$. Hence we have, as above,

$$\alpha_{Tf} = \inf\{\|Tf + h\|^2 : h \text{ in } A_2, \text{ Re } h(x_2) = 1 = \|h\|\}.$$

If g is in A_1 with $\text{Re } g(x_1) = 1 = \|g\|$, then $h = Tg$ is in A_2 with $\text{Re } h(x_2) = 1 = \|h\|$. Also, since T is onto, for every h in A_2 with $\text{Re } h(x_2) = 1 = \|h\|$, there is g in A_1 with $Tg = h$, so that $\text{Re } g(x_1) = \text{Re } h(x_2) = 1$ and $\|g\| = \|h\| = 1$. This shows that

$$\inf\{\|f + g\|^2 : g \text{ in } A_1, \text{ Re } g(x_1) = 1 = \|g\|\}$$
$$= \inf\{\|Tf + h\|^2 : h \text{ in } A_2, \text{ Re } h(x_2) = 1 = \|h\|\}.$$

Thus $\alpha_f = \alpha_{Tf}$; that is,

$$[\operatorname{Re} f(x_1) + 1]^2 + [\operatorname{Im} f(x_1)]^2 = [\operatorname{Re}(Tf)(x_2) + 1]^2 + [\operatorname{Im}(Tf)(x_2)]^2.$$

But $\operatorname{Re} f(x_1) = \operatorname{Re}(Tf)(x_2)$, by hypothesis. Hence $\operatorname{Im}(Tf)(x_2) = \pm \operatorname{Im} f(x_1)$ for all $f \in A_1$ with $\operatorname{Re} f(x_1) \geq 0$ and $\|f\| = 1$.

Now let f be an arbitrary nonzero element of A_1. Since $T(-f) = -f$ and $T(f/\|f\|) = Tf/\|f\|$, we can conclude from what we have proved above that $\operatorname{Im}(Tf)(x_2) = \pm \operatorname{Im} f(x_1)$.

In fact, we must have either $\operatorname{Im}(Tf)(x_2) = \operatorname{Im} f(x_1)$ for all f in A_1 or $\operatorname{Im}(Tf)(x_2) = -\operatorname{Im} f(x_1)$ for all f in A_1.

If this were false, there will exist f, g in A_1, such that $f(x_1) = a + ib$, $g(x_1) = c + id$,

$$T(f)(x_2) = a + ib \quad \text{and} \quad T(g)(x_2) = c - id \quad \text{with } b \neq 0, d \neq 0.$$

If $h := (f - a)/b + (g - c)/d$, then $h(x_1) = 2i$, whereas $(Th)(x_2) = 0$, a contradiction to what we have proved above.

Hence either $(Tf)(x_2) = f(x_1)$ for all f in A_1 or $(Tf)(x_2) = \overline{f(x_1)}$ for all f in A_1. $\square$

Notation Suppose that A_1 and A_2 are normed linear spaces and $T: A_1 \rightarrow A_2$ is a bounded linear map. Recall that A_1' and A_2' denote the dual spaces of A_1 and A_2, respectively. As usual, the *transpose of T* is the linear map $T': A_2' \rightarrow A_1'$ defined by $(T'\phi)(a) = \phi(T(a))$ for all ϕ in A_2' and a in A_1. Note that T' is a bounded linear map and $\|T'\| = \|T\|$ [see Limaye (1981), Theorem 13.9(a)].

Throughout the rest of the chapter, X_1 and X_2 are compact Hausdorff spaces, τ_1 and τ_2 are topological involutions on X_1 and X_2, respectively, and A_1, A_2 are (real) subspaces of $C(X_1, \tau_1)$ and $C(X_2, \tau_2)$, respectively, both containing 1.

Lemma 5.1.2 Let $\operatorname{Re} A_1$ separate the points of X_1/τ_1 and let T be a linear isometry of A_1 onto A_2 such that $T(1) = 1$. Then for every x_2 in $\operatorname{Ch}(A_2)$, there is some x_0 in $\operatorname{Ch}(A_1)$ such that

$$\operatorname{Re}(Tf)(x_2) = \operatorname{Re} f(x_0) \quad \text{for all } f \text{ in } A_1.$$

If, in addition, A_1 is extremely regular at $\{x_0, \tau_1(x_0)\}$ and A_2 is extremely regular at $\{x_2, \tau_2(x_2)\}$, then there is some x_1 in $\operatorname{Ch}(A_1)$ such that

$$(Tf)(x_2) = f(x_1) \quad \text{for all } f \text{ in } A_1.$$

Further, if A_1 separates x_0 and $\tau_1(x_0)$ whenever $x_0 \neq \tau_1(x_0)$, then x_1 is unique.

If A_2 also separates x_2 and $\tau_2(x_2)$ whenever $x_2 \neq \tau_2(x_2)$, then $x_2 = \tau_2(x_2)$ if and only if $x_1 = \tau_1(x_1)$.

Proof. Since T is a linear isometry of A_1 onto A_2, its transpose T' is a linear isometry of A_2' onto A_1'. Since $T(1) = 1$, it is easy to see that T' maps the set of all extreme points of $K(A_2)$ onto the set of all extreme points of $K(A_1)$.

Let $x_2 \in \mathrm{Ch}(A_2)$. By Theorem 4.1.10, $\mathrm{Re}(e_{x_2})$ is an extreme point of $K(A_2)$. Hence $T'(\mathrm{Re}(e_{x_2}))$ is an extreme point of $K(A_1)$. Since Re A_1 separates the points of X_1/τ_1, Theorem 4.1.10 shows that $T'(\mathrm{Re}(e_{x_2})) = \mathrm{Re}(e_{x_0})$ for some $x_0 \in \mathrm{Ch}(A_1)$; that is,

$$\mathrm{Re}(Tf)(x_2) = \mathrm{Re}(e_{x_2})(Tf)$$
$$= T'(\mathrm{Re}(e_{x_2}))(f) = \mathrm{Re}\, e_{x_0}(f)$$
$$= \mathrm{Re}\, f(x_0)$$

for all f in A_1, as desired.

If A_1 is extremely regular at $\{x_0, \tau_1(x_0)\}$ and A_2 is extremely regular at $\{x_2, \tau_2(x_2)\}$, then by Lemma 5.1.1, either $(Tf)(x_2) = f(x_0)$ for all f in A_1 or $(Tf)(x_2) = \overline{f(x_0)}$ for all f in A_1. In the former case, let $x_1 = x_0$, and in the latter, let $x_1 = \tau_1(x_0)$. Then $(Tf)(x_2) = f(x_1)$ for all f in A_1, where $x_1 \in \mathrm{Ch}(A_1)$ since $x_0 \in \mathrm{Ch}(A_1)$.

Now if for some $y_1 \in \mathrm{Ch}(A_1)$, $T(f)(x_2) = f(y_1)$ for all f in A_1, then Re $f(y_1) = \mathrm{Re}\, f(x_0) = \mathrm{Re}\, f(x_1)$ for all f in A_1. Since Re A_1 separates the points of X_1/τ_1, it follows that $\{x_1, y_1\} \subset \{x_0, \tau_1(x_0)\}$. Now, if $x_0 = \tau_1(x_0)$, then clearly $y_1 = x_0 = x_1$, and if $x_0 \neq \tau_1(x_0)$ and A_1 separates x_0 and $\tau_1(x_0)$, then again, $y_1 = x_1$. This implies the uniqueness of x_1.

Finally suppose further that if $x_2 \neq \tau_2(x_2)$, then A_2 separates x_2 and $\tau_2(x_2)$. Then $x_2 = \tau_2(x_2)$ if and only if $g(x_2)$ is real for all g in A_2 if and only if $(Tf)(x_2)$ is real for all f in A_1 (since T is onto) if and only if $f(x_1)$ is real for all f in A_1 if and only if $x_1 = \tau_1(x_1)$. $\square$

Lemma 5.1.3 Let Re A_1 separate the points of X_1/τ_1. Let A_1 separate x and $\tau_1(x)$ whenever $x \neq \tau_1(x)$ and $x \in \mathrm{Ch}(A_1)$. Let T be a linear isometry of A_1 onto A_2 with $T(1) = 1$, and assume that for $j = 1,2$, A_j is extremely regular at $\{x_j, \tau_j(x_j)\}$ for all x_j in $\mathrm{Ch}(A_j)$. Then there exists a continuous map $t: \mathrm{Ch}(A_2) \to \mathrm{Ch}(A_1)$ such that $\tau_1 \circ t = t \circ \tau_2$ and

$$(Tf)(x_2) = f(t(x_2)) \qquad \text{for all } f \in A_1,\ x_2 \in \mathrm{Ch}(A_2). \qquad (*)$$

If, in addition, Re A_2 separates the points of X_2/τ_2 and A_2 separates y and $\tau_2(y)$ whenever $y \neq \tau_2(y)$ and $y \in \mathrm{Ch}(A_2)$, then t is a homeomorphism.

Proof. The assumed separation conditions about A_1 imply that A_1 separates the points of $Ch(A_1)$. By Lemma 5.1.2, for each x_2 in $Ch(A_2)$, there exists a unique x_1 in $Ch(A_1)$ such that $(Tf)(x_2) = f(x_1)$ for all f in A_1. Defining $x_1 = t(x_2)$, we see that (*) is satisfied. Now suppose that a net $\{x_\alpha\}$ converges to x_2 in $Ch(A_2)$. Then from (*), $f(t(x_\alpha)) = (Tf)(x_\alpha)$ converges to $(Tf)(x_2) = f(t(x_2))$ for all f in A_1. As A_1 separates the points of $Ch(A_1)$, this implies that $\{t(x_\alpha)\}$ converges to $t(x_2)$. Hence t is continuous.

Again, using (*) we see that for all f in A_1 and x_2 in $Ch(A_2)$,

$$f(\tau_1(t(x_2))) = \bar{f}(t(x_2)) = (\overline{Tf})(x_2) = (Tf)(\tau_2(x_2))$$
$$= f(t(\tau_2(x_2))).$$

Since A_1 separates the points of $Ch(A_1)$, we have $\tau_1(t(x_2)) = t(\tau_2(x_2))$ for all x_2 in $Ch(A_2)$; that is, $\tau_1 \circ t = t \circ \tau_2$.

Next, if A_2 also satisfies similar separation conditions, then applying the same argument to the isometry $T^{-1}: A_2 \to A_1$ [note that $T^{-1}(1) = 1$] we get a continuous map $s: Ch(A_1) \to Ch(A_2)$ such that

$$(T^{-1}g)(x_1) = g(s(x_1)) \qquad \text{for all } g \text{ in } A_2 \text{ and } x_1 \text{ in } Ch(A_1).$$

Hence for all f in A_1, we have

$$f(x_1) = (T^{-1}Tf)(x_1) = (Tf)(s(x_1)) = f(t(s(x_1))).$$

Since A_1 separates the points of $Ch(A_1)$, we have $x_1 = t(s(x_1))$ for all x_1 in $Ch(A_1)$. Similarly, we can prove that $s(t(x_2)) = x_2$ for all x_2 in $Ch(A_2)$. Thus t has a continuous inverse and is hence a homeomorphism. $\qquad\square$

Theorem 5.1.4 For $j = 1,2$, let A_j be a real function algebra on (X_j, τ_j). Let T be a linear map of A_1 onto A_2. Then the following statements are equivalent:
(i) T is an isometry and $T(1) = 1$.
(ii) There is a homeomorphism $t: Ch(A_2) \to Ch(A_1)$ such that $\tau_1 \circ t = t \circ \tau_2$ and $(Tf)(x) = f(t(x))$ for all x in $Ch(A_2)$ and f in A_1.
(iii) T is an algebra isomorphism.

Proof.
(i) implies (ii): For $j = 1,2$, A_j separates the points of X_j by definition, $Re\ A_j$ separates the points of X_j/τ_j by Lemma 1.3.9 and A_j is extremely regular at $\{x_j, \tau_j(x_j)\}$ for each x_j in $Ch(A_j)$ by Theorem 4.2.4(v). Now (ii) follows from Lemma 5.1.3.

(ii) implies (iii): Let $f, g \in A_1$ and $x_2 \in Ch(A_2)$. Then

$$T(fg)(x_2) = (fg)(t(x_2)) = f(t(x_2))g(t(x_2)) = (Tf)(x_2)(Tg)(x_2).$$

Since this is true for all x_2 in $Ch(A_2)$, and $Ch(A_2)$ is a boundary by

Theorem 4.2.5, the equation holds for all x_2 in X_2. Thus $T(fg) = T(f)T(g)$ for all f,g in A_1.

(iii) implies (i): Let $T\colon A_1 \to A_2$ be an algebra isomorphism. Then T is linear, and clearly $T(1) = 1$. Further, it follows from Remark 1.1.22 and Corollary 1.2.12 that $\|Tf\| = \|f\|$ for every f in A_1; that is, T is an isometry.   □

Remark 5.1.5 Let X_1 and X_2 be compact Hausdorff spaces and B_1 and B_2 complex function algebras on X_1 and X_2, respectively. Let $T\colon B_1 \to B_2$ be a (complex) linear isometry such that $T(1) = 1$. Since B_1 and B_2 can be regarded as real function algebras (Remark 1.3.19), Theorem 5.1.4 implies that T is an isomorphism of the real algebra B_1 to the real algebra B_2. But since T is already complex linear, it is a complex algebra isomorphism. Thus we get the classical theorem of Nagasawa.

Remark 5.1.6 The above-mentioned result of Nagasawa (in which units are preserved by the isometry) for complex function algebras has been generalized further by Jarosz (1983) as follows: "Let B_1 be a complex function algebra with unit e_1, and B_2 be a complex Banach algebra with unit e_2. Suppose that T is a linear isometry of B_1 onto B_2 such that $T(e_1) = e_2$. Then T is an algebra isomorphism." However, this result is not valid for real function algebras, as the counterexample given by Jarosz (1983) and quoted in Example 1.4.6 shows: For $0 \le t \le \frac{1}{2}$, each A_t is a commutative Banach algebra with unit $(1,1)$. A_0 is $\mathbb{R}^2$ with the usual multiplication. Hence A_0 is isometrically isomorphic to $C_\mathbb{R}(\{x_1,x_2\})$. Thus A_0 is a real function algebra. The identity map of A_0 onto A_t for $0 < t \le \frac{1}{2}$ is a linear isometry of A_0 onto A_t, which preserves units, but it is not an isomorphism.

Remark 5.1.7 Mazur and Ulam (1932) have proved that an isometry of a real Banach space onto a real Banach space which preserves the origin is linear. In view of this result, the statement of Theorem 5.1.4 can be strengthened as follows: A mapping T from A_1 onto A_2 is an algebra isomorphism if and only if T is an isometry of A_1 onto A_2 such that $T(0) = 0$ and $T(1) = 1$.

5.2 ARBITRARY ISOMETRIES

If there exists a linear isometry between two complex function algebras, then it is possible to find one that preserves units [Lemma 30.2 of Zelazko (1973)]. To establish this in the case of real function algebras, we need a few lemmas.

Lemma 5.2.1 Let X be a compact Hausdorff space, τ be a topological involution on X, A be a real function algebra on (X,τ), and $f \in A$ with $\|f\| = 1$. Then $|f| \equiv 1$ on $Ch(A)$ if and only if for every $\varepsilon > 0$, there exists $\delta > 0$ such that for every g in A with $\|g\| \geq \varepsilon$, we have

$$\max\{\|f + g\|, \quad \|f - g\|\} > 1 + \delta.$$

Proof. Assume that $|f| \equiv 1$ on $Ch(A)$. Let $\varepsilon > 0$ and g in A be such that $\|g\| \geq \varepsilon$. Then $|g(x)| \geq \varepsilon$ for some x in $Ch(A)$, as $Ch(A)$ is a boundary (Theorem 4.2.5). Since $|f(x)| = 1$ and

$$|f(x) + g(x)|^2 + |f(x) - g(x)|^2 = 2|f(x)|^2 + 2|g(x)|^2 \geq 2 + 2\varepsilon^2,$$

we see that

$$\max\{|f(x) + g(x)|^2, \quad |f(x) - g(x)|^2\} \geq 1 + \varepsilon^2.$$

Hence

$$\max\{\|f + g\|, \quad \|f - g\|\} > \sqrt{1 + \varepsilon^2} > 1 + \delta,$$

if $\delta < \sqrt{1 + \varepsilon^2} - 1$.

Conversely, assume that there exists x_0 in $Ch(A)$ such that $|f(x_0)| < 1$. We can find $\varepsilon > 0$ such that $|f(x_0)| \leq 1 - 2\varepsilon$. Then $V := \{x \in X : |f(x)| < 1 - \varepsilon\}$ is a τ-invariant neighborhood of $\{x_0, \tau(x_0)\}$. Consider $\delta > 0$. Now by Theorem 4.2.4, A is extremely regular at $\{x_0, \tau(x_0)\}$. Hence there exists g in A such that $\|g\| = \varepsilon = g(x_0)$ and $|g(y)| < \delta$ for all y in $X \setminus V$. Then for x in V, $|f(x) \pm g(x)| \leq 1 - \varepsilon + \varepsilon = 1$, and for x in $X \setminus V$, $|f(x) \pm g(x)| \leq 1 + \delta$. Hence $\max\{\|f + g\|, \|f - g\|\} \leq 1 + \delta$. $\qquad\square$

Lemma 5.2.2 Let for $j = 1,2$, A_j be a real function algebra on (X_j, τ_j). Suppose that $T : A_1 \to A_2$ is a linear isometry of A_1 onto A_2 and $T(1) = h$. Then
(i) $|h| \equiv 1$ on $Ch(A_2)$.
(ii) h is invertible in A_2.
(iii) The map $T_1 : A_1 \to A_2$ defined by

$$T_1(f) := h^{-1}T(f), \qquad f \in A_1,$$

is a linear isometry of A_1 onto A_2 such that $T_1(1) = 1$.

Proof.
(i) Let $\varepsilon > 0$. By Lemma 5.2.1 there exists $\delta > 0$ such that for all g_1 in A_1 with $\|g_1\| \geq \varepsilon$, we have

$$\max\{\|1 + g_1\|, \quad \|1 - g_1\|\} > 1 + \delta.$$

Now, let $g_2 \in A_2$ with $\|g_2\| \geq \varepsilon$. Since T is onto, there is g_1 in A_1 such that

$Tg_1 = g_2$. Then since T is an isometry, $\|g_1\| = \|g_2\| \geq \varepsilon$, and

$$\max\{\|h + g_2\|, \quad \|h - g_2\|\} = \max\{\|T(1 + g_1)\|, \quad \|T(1 - g_1)\|\}$$
$$= \max\{\|1 + g_1\|, \quad \|1 - g_1\|\} > 1 + \delta.$$

Hence, again by Lemma 5.2.1, $|h| \equiv 1$ on $\mathrm{Ch}(A_2)$.

(ii) Let $S_2 = S(A_2)$ be the Shilov boundary of A_2. Since S_2 is the closure of $\mathrm{Ch}(A_2)$, [Definition 4.2.6] we have $|h| \equiv 1$ on S_2. Now, we define

$$A_0 := \{(\bar{h}f)|_{S_2} : f \in A_2\}.$$

It is easy to see that A_0 is a uniformly closed subspace of $C(S_2, \tau_{2|S_2})$, it separates the points of S_2 and contains real constants.

Let $g \in A_2$. Then $hg \in A_2$. Since $|h| \equiv 1$ on S_2, we have $g|_{S_2} = (\bar{h}hg)|_{S_2} \in A_0$. Thus $A_2|_{S_2} \subset A_0$. Since A_2 and $A_2|_{S_2}$ are isometrically isomorphic, we may regard A_2 as a subset of A_0.

Define a map $T_0: A_1 \to A_0$ by

$$T_0(f) := (\bar{h}T(f))|_{S_2} \qquad \text{for } f \text{ in } A_1.$$

Then T_0 is a linear isometry of A_1 onto A_0 and $T_0(1) = 1$.

Let $x_2 \in \mathrm{Ch}(A_2)$. By Theorem 4.2.4, A_2 is extremely regular at $\{x_2, \tau_2(x_2)\}$, and so is A_0, because $A_2 \subset A_0$. Hence $x_2 \in \mathrm{Ch}(A_0)$ by Corollary 4.1.7. Now by Lemma 5.1.2, there is a unique x_1 in $\mathrm{Ch}(A_1)$ such that $T_0(f)(x_2) = f(x_1)$ for all f in A_1; that is,

$$(\bar{h}T(f))(x_2) = f(x_1) \qquad \text{for all } f \text{ in } A_1.$$

Since T is onto, there is f_1 in A_1 such that $T(f_1) = 1$. Then

$$\bar{h}(x_2) = f_1(x_1) \qquad \text{and} \qquad (\bar{h}T(f_1^2))(x_2) = f_1^2(x_1),$$

so that $(hT(f_1^2) - 1)(x_2) = 0$.

This equation holds for all x_2 in $\mathrm{Ch}(A_2)$ and hence for all x_2 in X_2 as $\mathrm{Ch}(A_2)$ is a boundary for A_2 (Theorem 4.2.5) and $(hT(f_1^2) - 1) \in A_2$. This shows that $T(f_1^2)$ is the inverse of h in A_2.

(iii) This follows from (i) and (ii). $\qquad\qquad\qquad\qquad\qquad\qquad\square$

We are now in a position to prove the following analog of Nagasawa's theorem.

Theorem 5.2.3 Let for $j = 1,2$, A_j be a real function algebra on (X_j, τ_j). Then A_1 and A_2 are isomorphic if and only if they are linearly isometric.

Proof. Let $T: A_1 \to A_2$ be a linear isometry of A_1 onto A_2. Then by

Lemma 5.2.2, there exists a linear isometry T_1 of A_1 onto A_2 such that $T_1(1) = 1$. Hence by Theorem 5.1.4, T_1 is an algebra isomorphism, showing that A_1 and A_2 are algebraically isomorphic.

Conversely, let $T: A_1 \to A_2$ be an algebra isomorphism. Then T is a linear isometry of A_1 onto A_2 as in the proof of Theorem 5.1.4. $\square$

The proofs of Lemmas 5.1.1 and 5.1.2, Theorem 5.1.4, and Lemma 5.2.2 are modifications of proofs in Kulkarni and Arundhathi (1991a). Grzesiak (1990) has proved the following extension of the Banach–Stone theorem.

Corollary 5.2.4 A map $T: C(X_1, \tau_1) \to C(X_2, \tau_2)$ is a linear onto isometry if and only if there exists a homeomorphism $t: X_2 \to X_1$ with $\tau_1 \circ t = t \circ \tau_2$ and a function $h \in C(X_2, \tau_2)$ with $|h| \equiv 1$ on X_2 such that

$$(Tf)(x_2) = h(x_2)f(t(x_2)) \qquad \text{for every } f \text{ in } C(X_1, \tau_1) \text{ and } x_2 \text{ in } X_2.$$

Proof. Let T be a linear onto isometry. Note that $\text{Ch}(C(X_j, \tau_j)) = X_j$ for $j = 1, 2$ by Theorem 4.1.3. Let $T(1) = h$. By Lemma 5.2.2, $|h| \equiv 1$ on X_2 and $T_1(f) = h^{-1}T(f)$, $f \in C(X_1, \tau_1)$, defines a linear isometry of $C(X_1, \tau_1)$ onto $C(X_2, \tau_2)$ with $T_1(1) = 1$. Now the conclusion follows from Theorem 5.1.4.

The converse follows since for $g \in C(X_2, \tau_2)$, $T((\bar{h}g) \circ t^{-1}) = g$. $\square$

Symbols

References

1. N. L. Alling (1970), Real Banach algebras and nonorientable Klein sur-faces—I, J. Reine Angew. Math. *241*, 200–208. MR *41*, 5972.
2. N. L. Alling and L. A. Campbell (1972), Real Banach algebras—II, Math. Z. *125*, 79–100. MR *45*, 4150.
3. N. L. Alling and N. Greenleaf (1971), Foundations of the Theory of Klein Surfaces, Springer-Verlag, Berlin. MR *48*, 11488.
4. N. L. Alling and B. V. Limaye (1972), Ideal theory on non-orientable Klein surfaces, Ark. Mat. *10*, 277–292. MR *53*, 3892.
5. R. Arens and I. Kaplansky (1948), Topological representation of alge-bras, Trans. Amer. Math. Soc. *63*, 457–481. MR *11*, 317.
6. S. Arundhathi (1988), Studies in real function algebras, Ph.D. thesis, Indian Institute of Technology, Madras, India.
7. S. Arundhathi (1990), A note on the peak points for real function algebras, Indian J. Pure Appl. Math. *21*, 155–162. MR *91f*, 46071.
8. S. Arundhathi and S. H. Kulkarni (1986), Analytic and harmonic maps into a topological space, Proc. Indian Acad. Sci. (Math. Sci.) *95*, 37–40. MR *88h*, 46096.
9. H. S. Bear (1965), A geometric characterization of Gleason parts, Proc. Amer. Math. Soc. *16*, 407–412. MR *31*, 5136.
10. H. S. Bear (1970), Lectures on Gleason parts, Lecture Notes in Mathe-matics 121, Springer-Verlag, Berlin. MR *41*, 5950.
11. E. Behrends (1979), *M*-Structure and the Banach–Stone theorem, Lecture Notes in Mathematics 736, Springer-Verlag, Berlin. MR *81b*, 46002.

177

12. A. Bernard (1968), Une characterization de $C(X)$ paemi les algebres de Banach, C. R. Acad. Sci. Paris *267*, 634–635. MR *38*, 2601.

13. A. Bernard (1970), Fonctions qui opérent sur Re A, C. R. Acad. Sci. Paris *271*, 1120–1122. MR *42*, 5048.

14. A. Bernard (1972), Espace des parties réelles des éléments d'une algèbre de Banach de fonctions, J. Funct. Anal. *10*, 387–409. MR *49*, 7781.

15. E. Bishop (1961), A generalization of the Stone–Weierstrass theorem, Pacific J. Math. *11*, 777–783. MR *24*, A 3502.

16. F. F. Bonsall and J. Duncan (1973), Complete Normed Algebras, Springer-Verlag, Berlin. MR *54*, 11013.

17. L. de Branges (1959), The Stone–Weierstrass theorem, Proc. Amer. Math. Soc. *10*, 822–824. MR *22*, 3970.

18. B. Brosowski and F. Deutsch (1981), An elementary proof of the Stone–Weierstrass theorem, Proc. Amer. Math. Soc. *81*, 89–92. MR *81j*, 46086.

19. A. Browder (1969), Introduction to Function Algebras, W. A. Benjamin, New York. MR *39*, 7431.

20. R. B. Burckel (1972), Characterizations of $C(X)$ among its subalgebras, Lecture Notes in Pure and Applied Mathematics, Vol. 6, Marcel Dekker, New York. MR *56*, 1068.

21. R. B. Burckel (1984), Bishop's Stone-Weierstrass theorem, Amer. Math. Monthly. *91*, 22–32. MR *85i*, 46071.

22. J. Cerych (1986), A remark on linear codimensions of function algebras, Czechoslovak Math. J. *36*, 511–515. MR *87j*, 46097.

23. T. W. Gamelin (1969), Uniform Algebras, Prentice Hall, Englewood Cliffs, N.J. MR *53*, 14137.

24. J. Garnett (1967), A topological characterization of Gleason parts, Pacific J. Math. *20*, 59–63. MR *34*, 4942.

25. I. M. Gelfand (1941). Normierte ringe, Mat. Sbornik N.S. *9*(51), 3–24. MR *3*, 4486.

26. A. M. Gleason (1957), Function algebras, Seminars on Analytic Functions, Institute for Advanced Study, Princeton, N.J.

27. I. Glicksberg (1962), Measures orthogonal to algebras and sets of anti-symmetry, Trans. Amer. Math. Soc. *105*, 415–435. MR *30*, 4164.

28. K. R. Goodearl (1982), Notes on Real and Complex C^*-Algebras, Shiva Publishing, Nantwich, Cheshire, England. MR *85d*, 46079.

29. E. A. Gorin (1965), Moduli of invertible elements in a normed algebra, Vestnik Moskov. Univ. Ser. I Mat. Meh. *5*, 35–39. MR *32*, 8206.

30. M. Grzesiak (1986), Extreme points of the unit ball in the dual space of some real Banach algebra of continuous complex functions, Fasc. Math. *16*, 5–10. MR *88c*, 46062.

31. M. Grzesiak (1989), Real function algebras and their sets of antisymmetry, Glasnik Mat. *24*, 297–304. MR *91i*, 46055.

32. M. Grzesiak (1990), Isometries of a space of continuous functions determined by an involution, Math. Nachr. *145*, 217–221.

33. M. Grzesiak (1991), Localising families for real function algebras (to appear).

34. O. Hatori (1981), Functions which operate on the real part of a function algebra, Proc. Amer. Math. Soc. *83*, 565–568. MR *82j*, 46063.

35. R. A. Hirschfeld and W. Zelazko (1968), On spectral norm Banach algebras, Bull. Acad. Polon. Sci. *16*, 195–199. MR *37*, 4621.

36. K. Hoffman and J. Wermer (1962), A characterization of $C(X)$, Pacific J. Math. *12*, 941–944. MR *27*, 325.

37. S. Hwang (1990), Aspects of commutative Banach algebras, Ph.D. thesis, University of Connecticut.

38. L. Ingelstam (1962), A vertex property for Banach algebras with identity, Math. Scand. *2*, 22–32. MR *26*, 4199.

39. L. Ingelstam (1963), Hilbert algebras with identity, Bull. Amer. Math. Soc. *69*, 794–796. MR *27*, 4096.

40. L. Ingelstam (1964), Real Banach algebras, Ark. Mat. *5*, 239–270. MR *30*, 2358.

41. L. Ingelstam (1968), Symmetry in real Banach algebras, Math. Scand. *18*, 53–68. MR *34*, 6555.

42. L. Ingelstam (1967), A note on Laplace transforms and strict reality in Banach algebras, Math. Z. *102*, 163–165. MR *36*, 6936.

43. L. Ingelstam (1969). On semigroups generated by topological nilpotent elements, Illinois J. Math. *13*, 172–175. MR *38*, 4996.

44. K. Jarosz (1983), The uniqueness of multiplication in function algebras, Proc. Amer. Math. Soc. *89*, 249–253. MR *85c*, 46048.

45. K. Jarosz (1984a), A characterization of weak peak sets for function algebras, Bull. Austral. Math. Soc. *29*, 129–135. MR *86a*, 46058.

46. K. Jarosz (1984b), Into isomorphisms of spaces of continuous functions, Proc. Amer. Math. Soc. *90*, 373–377. MR *85k*, 46024.

47. K. Jarosz (1985), Perturbations of Banach algebras, Lecture Notes in Mathematics 1120, Springer-Verlag, Berlin. MR *86k*, 46074.

48. K. Jarosz and Z. Sawon (1985), A discontinuous function does not operate on the real part of a function algebra, Casopis Pest. Mat. Soc. *110*, 58–59. MR *86g*, 46080.

49. B. E. Johnson (1977), Perturbations of Banach algebras, Proc. London Math. Soc. *34*, 439–458. MR *56*, 1094.

50. R. V. Kadison (1951), A representation theory for a commutative topological algebra, Mem. Amer. Math. Soc. *7*, 39–50. MR *13*, 360.

51. R. V. Kadison and D. Kastler (1972), Perturbations of von Newmann algebra I—stability of type, Amer. J. Math. *94*, 38–54. MR *45*, 5772.

52. I. Kaplansky (1949), Normed algebras, Duke Math. J. *16*, 399–418. MR *11*, 115.

53. H. Konig (1966), Zur abstracten Theorie analytischen Funktionen II, Math. Ann. *163*, 9–17. MR *32*, 8202.

54. M. König (1969), On the Gleason and Harnak metrics for uniform algebras, Proc. Amer. Math. Soc. *22*, 100–101. MR *39*, 3313.

55. S. H. Kulkarni (1983), Analytic and harmonic maps into the maximal ideal space of a function algebra, J. Math. Phys. Sci. *17*, 169–175. MR *84j*, 46081.

56. S. H. Kulkarni (1988), Topological conditions for commutativity of a real Banach algebra, Houston J. Math. *14*, 235–245. MR *90b*, 46069.

57. S. H. Kulkarni and S. Arundhathi (1988), Choquet boundary for real function algebras, Canad. J. Math. *40*, 1084–1104. MR *90a*, 46129.

58. S. H. Kulkarni and S. Arundhathi (1991a), Isometries of real function algebras, Comment. Math. *30*, 343–356.

59. S. H. Kulkarni and S. Arundhathi (1991b), A note on the metric topology on Gleason parts of a real function algebra (to appear).

60. S. H. Kulkarni and S. Arundhathi (1991c), Perturbations of real function algebras (to appear).

61. S. H. Kulkarni and B. V. Limaye (1980), Gelfand–Naimark theorems for real Banach algebras, Math. Japon. *25*, 545–558. MR *82c*, 46071.

62. S. H. Kulkarni and B. V. Limaye (1981a), Spectral mapping theorem for real Banach algebras, Houston J. Math. *7*, 507–517. MR *83i*, 46057.

63. S. H. Kulkarni and B. V. Limaye (1981b), Gleason parts of real function algebras, Canad. J. Math. *33*, 181–200. MR *82m*, 46052.

64. S. H. Kulkarni and B. V. Limaye (1983), A topological characterization of Gleason parts of real function algebras, Canad. Math. Bull. *26*, 44–49. MR *83m*, 46076.

65. S. H. Kulkarni and B. V. Limaye (1991a), Choquet sets and boundaries for real subspaces of $C(X)$ (to appear).

66. S. H. Kulkarni and B. V. Limaye (1991b), Extreme points of the unit ball of the dual space of a subspace of $C(X)$ (to appear).

67. S. H. Kulkarni and N. Srinivasan (1987), An analogue of Bishop's theorem for real function algebras, Indian J. Pure Appl. Math. *18*, 136–145. MR *83d*, 46101.

68. S. H. Kulkarni and N. Srinivasan (1988a), An analogue of Hoffman–Wermer theorem for real function algebras, Indian J. Pure Appl. Math. *19*, 154–166. MR *89i*, 46059.

69. S. H. Kulkarni and N. Srinivasan (1988b), A note on a theorem of Gorin, Math. Japon. *33*, 887–893. MR *90a*, 46130.

70. S. H. Kulkarni and N. Srinivasan (1990), An analogue of Wermer's theorem for a real function algebra, Math. Today *8*, 33–42.

71. K. de Leeuw and Y. Katznelson (1963), Functions that operate on non-selfadjoint algebras, J. Analyse Math. *11*, 207–219. MR *28*, 1508.

72. B. V. Limaye (1976), Boundaries for real Banach algebras, Canad. J. Math. *28*, 42–49. MR *52*, 15022.

73. B. V. Limaye (1981), Functional Analysis, Wiley Eastern, New Delhi. MR *83b*, 46001.

74. B. V. Limaye, R. D. Mehta, and M. H. Vasavada (1982), Maximal ideal space and Šilov boundary of the tensor product of real Banach algebras, Glas. Mat. *17*, 277–283. MR *84h*, 46069.

75. B. V. Limaye and R. R. Simha (1975), Deficiencies of certain real uniform algebras, Canad. J. Math. *27*, 121-132. MR *52*, 3962.
76. S. Machado (1977), On Bishop's generalization of the Stone-Weierstrass theorem, Nederl. Akad. Wetensch. Proc. Ser. A 80, Indag. Math. *39*, 218-224. MR *56*, 6356.
77. S. Mazur and S. Ulam (1932), Sur les transformations isométriques d'espace vectoriels normes, C. R. Acad. Sci. Paris, *194*, 946-948.
78. M. S. Mehta, R. D. Mehta, and M. H. Vasavada (1990), Sĭlov and other decompositions for a real function algebra, Math. Today *8*, 1-12.
79. R. D. Mehta and M. H. Vasavada (1986), Wermer's type result for a real Banach function algebra, Math. Today *4*, 43-46. MR *86c*, 46064.
80. M. Nagasawa (1959), Isomorphisms between commutative Banach algebras with an application to rings of analytic functions, Kodai Math. Sem. Rep. *11*, 182-188. MR *22*, 12379.
81. H. W. Oliver (1970), Noncomplex methods in real Banach algebras, J. Funct. Anal. *6*, 401-411. MR *42*, 3581.
82. T. Ono (1970), A real analogue of the Gelfand-Naimark theorem, Proc. Amer. Math. Soc. *25*, 159-160. MR *41*, 2407.
83. C. Le Page (1967), Sur quelques conditions entraînant la commutativite dans les algèbres des Banach, C. R. Acad. Sci. Paris Ser. A-B *265*, 235-237. MR *37*, 1999.
84. T. W. Palmer (1972), Real C^* algebras, Pacific J. Math. *6*, 245-290.
85. R. R. Phelps (1966), Lectures on Choquet's theorem. D. Van Nostrand, Princeton, N.J. MR *33*, 1690.
86. J. Phillips (1973), Perturbations of C^* algebras, Indiana Univ. Math. J. *23*, 1167-1176. MR *49*, 5861.
87. J. B. Prolla (1971), Bishop's generalized Stone-Weierstrass theorem for weighted spaces, Math. Ann. *191*, 283-289. MR *44*, 7200.
88. J. B. Prolla (1988), A generalized Bernstein approximation theorem, Math. Proc. Cambridge Philos. Soc. *104*, 317-330. MR *90b*, 41055.
89. I. Raeburn and J. L. Taylor (1977), Hochschild cohomology and perturbations of Banach algebras, J. Funct. Anal. *25*, 258-267. MR *58*, 30334.
90. T. J. Ransford (1984), A short elementary proof of the Bishop-Stone-Weierstrass theorem, Math. Proc. Cambridge Philos. Soc. *96*, 309-311. MR *86c*, 46023.
91. C. E. Rickart (1960), General Theory of Banach Algebras, D. Van Nostrand, Princeton, N.J., MR *22*, 5903.
92. R. Rochberg (1979), Deformation of uniform algebras, Proc. London Math. Soc. *39*, 93-118. MR *80k*, 46058.
93. W. Rudin (1957), The closed ideals in an algebra of analytic functions, Canad. J. Math. *9*, 426-434. MR *19*, 641.
94. W. Rudin (1964), Principles of Mathematical Analysis, McGraw-Hill, New York. MR *29*, 3587.
95. W. Rudin (1966), Real and Complex Analysis, McGraw-Hill, New York. MR *35*, 1420.

96. W. Rudin (1973), Functional Analysis, McGraw-Hill, New York. MR *51*, 1315.

97. Z. Semadeni (1971), Banach Spaces of Continuous Functions, PWN—Polish Scientific Publishers, Warsaw. MR *45*, 5930.

98. G. E. Shilov (1951), On rings of functions with uniform convergence, Ukrain. Mat. Z. *3*, 404–411. MR *14*, 884.

99. S. J. Sidney (1979), Functions which operate on a real part of a uniform algebra, Pacific J. Math. *80*, 265–272. MR *81b*, 46069.

100. S. J. Sidney and E. L. Stout (1968), A note on interpolation, Proc. Amer. Math. Soc. *19*, 380–382. MR *81b*, 46069.

101. G. F. Simmons (1963), Introduction to Topology and Modern Analysis, McGraw-Hill, New York. MR *26*, 4145.

102. G. Springer (1957), *Introduction to Riemann Surfaces*, Addison-Wesley, Reading, Massachusetts. MR *19*, 1169.

103. N. Srinivasan (1988), Characterizations of $C(X,\tau)$ among its subalgebras, Ph.D. thesis, Indian Institute of Technology, Madras, India.

104. N. Srinivasan and S. H. Kulkarni (1988), Restriction algebras of a real function algebra, J. Math. Phys. Sci. *22*, 209–223. MR *89k*, 46087.

105. A. Srivastav (1990), Commutativity criterion for real Banach algebras, Arch. Math. *54*, 65–72. MR *91b*, 46044.

106. A. Srivastav (1991), Extreme points of positive functionals and spectral states on real Banach algebras (to appear).

107. E. L. Stout (1971), The Theory of Uniform Algebras, Bogden and Quigley, Tarrytown-on-Hudson, N.Y. MR *54*, 11066.

108. I. Suciu (1975), Function Algebras, Noordhoff International Publishing, Leyden, The Netherlands. MR *51*, 6248.

109. J. Wermer (1963), The space of real parts of a function algebra, Pacific J. Math. *13*, 1423–1426. MR *27*, 6152.

110. W. Zelazko (1973), Banach algebras PWN—Polish Scientific Publishers, Warsaw. MR *56*, 6389.

Index